GETTING STARTED WITH STATA
FOR UNIX®
RELEASE 11

A Stata Press Publication
StataCorp LP
College Station, Texas

Table of Contents

Cross-referencing the documentation

When reading this manual, you will find references to other Stata manuals. For example,

[U] **26 Overview of Stata estimation commands**

[R] **regress**

[D] **reshape**

The first example is a reference to chapter 26, *Overview of Stata estimation commands*, in the *User's Guide*; the second is a reference to the `regress` entry in the *Base Reference Manual*; and the third is a reference to the `reshape` entry in the *Data-Management Reference Manual*.

All the manuals in the Stata Documentation have a shorthand notation:

[GSM]	*Getting Started with Stata for Mac*
[GSU]	*Getting Started with Stata for Unix*
[GSW]	*Getting Started with Stata for Windows*
[U]	*Stata User's Guide*
[R]	*Stata Base Reference Manual*
[D]	*Stata Data-Management Reference Manual*
[G]	*Stata Graphics Reference Manual*
[XT]	*Stata Longitudinal-Data/Panel-Data Reference Manual*
[MI]	*Stata Multiple-Imputation Reference Manual*
[MV]	*Stata Multivariate Statistics Reference Manual*
[P]	*Stata Programming Reference Manual*
[SVY]	*Stata Survey Data Reference Manual*
[ST]	*Stata Survival Analysis and Epidemiological Tables Reference Manual*
[TS]	*Stata Time-Series Reference Manual*
[I]	*Stata Quick Reference and Index*
[M]	*Mata Reference Manual*

Detailed information about each of these manuals may be found online at

http://www.stata-press.com/manuals/

About this manual

This manual discusses **Stata for Unix**®. Stata for Windows® users should see *Getting Started with Stata for Windows*; Stata for Mac® users should see *Getting Started with Stata for Mac*. This manual is intended both for people completely new to Stata and for experienced Stata users new to Stata for Unix. Previous Stata users will also find it helpful as a tutorial on some new features in Stata for Unix and as a reference for Unix-specific Stata commands.

Following the numbered chapters are three appendices with information specific to Stata for Unix and a fourth appendix with three sections devoted to Unix-specific commands in Stata.

We provide several types of technical support to registered Stata users. [GSU] **4 Getting help** describes the resources available to help you learn about Stata's commands and features. One of these resources is the Stata web site (http://www.stata.com), where you will find answers to frequently asked questions (FAQs), as well as other useful information. If you still have questions after looking at the Stata web site and the other resources described in [GSU] **19 Updating and extending Stata—Internet functionality**, you can contact us as described in [U] **3.9 Technical support**.

Using this manual

The new user will get the most out of this book by treating it as an exercise book, working through each example at the computer. The material builds, so material from earlier chapters will often be used in later chapters. Bear in mind that Stata is a rich and deep statistical package—just like statistics itself. The time spent working the examples will be repaid with dividends when doing true statistical analyses.

The experienced user may still have something to learn from this manual, despite its name. We suggest looking through the chapters to see if there is anything new or forgotten.

Stata's two interfaces

With Stata for Unix, you can choose between two user interfaces. The first option is the graphical user interface, or GUI, which we will refer to as Stata(GUI). The second option is the nongraphical user interface, which we will refer to as Stata(console). If instructions apply to either interface, we will refer simply to Stata.

This manual shows how to complete various tasks by using both the GUI and the console. When the manual states that a command can be typed into the Command window, we imply that the command could also be typed at the prompt in the console.

1 Introducing Stata—sample session

Introducing Stata

This chapter will run through a sample work session, introducing you to a few of the basic tasks that can be done in Stata, such as opening a dataset, investigating the contents of the dataset, using some descriptive statistics, making some graphs, and doing a simple regression analysis. As you would expect, we will only brush the surface of many of these topics. This approach should give you a sample of what Stata can do and how Stata works. There will be brief explanations along the way, with references to chapters later in this book, as well as to the online help and other Stata manuals. We will run through the session by using both menus and dialogs and Stata's commands, so that you can gain some familiarity with both.

If you are using Stata(console) rather than Stata(GUI) under Unix, you will not be able to use the menus and dialogs as they are mentioned here. However, the commands used to produce the output are included in the output in each case, so you can still work through the examples.

Take a seat at your computer, put on some good music, and work along with the book.

Sample session

The dataset that we will use for this session is a set of data about vintage 1978 automobiles sold in the United States.

To follow along using point-and-click, note that the menu items are given by **Menu > Menu Item > Submenu Item > etc**. To follow along using the Command window, type the commands that follow a dot (.) in the boxed listings below into the small window labeled **Command**. When there is something of note about the structure of a command, it will be pointed out as a "Syntax note".

Start by loading the auto dataset, which is included with Stata. To use the menus,
1. Select **File > Example Datasets...**.
2. Click on Example datasets installed with Stata.
3. Click on use for auto.dta.
The result of this command is threefold:
  • Some output appears in the large Results window.

```
. sysuse auto
(1978 Automobile Data)
```

The output consists of a command and its result. The command is bold and follows the period (.): sysuse auto. The output is in the standard face here and is a brief description of the dataset.

Note: If a command intrigues you, you can type help *commandname* in the Command window to find help. If you want to explore at any time, **Help > Search...** can be informative.

1

- The same command, `sysuse auto`, appears in the small Review window to the upper left. The Review window keeps track of commands Stata has run, successful and unsuccessful. The commands can then easily be rerun. See [GSU] **2 The Stata user interface** for more information.
- A series of variables appears in the small Variables window to the lower left.

You could have opened the dataset by typing `sysuse auto` in the Command window and pressing *Enter*. Try this now. `sysuse` is a command that loads (uses) example (system) datasets. As you will see during this session, Stata commands are often simple enough that it is faster to use them directly. This will be especially true once you become familiar with the commands you use the most in your daily use of Stata.

Syntax note: In the above example, `sysuse` is the Stata command, whereas `auto` is the name of a Stata data file.

Simple data management

We can get a quick glimpse at the data by browsing it in the **Data Editor**. This can be done by clicking on the **Data Editor (Browse)** button, 🔍 , or by selecting **Data > Data Editor > Data Editor (Browse)** from the menus, or typing the command `browse`.

Syntax note: Here the command is `browse` and there are no other arguments.

When the Data Editor window opens, you can see that Stata regards the data as one rectangular table. This is true for all Stata datasets. The columns represent *variables*, whereas the rows represent *observations*. The variables have somewhat descriptive names, whereas the observations are numbered.

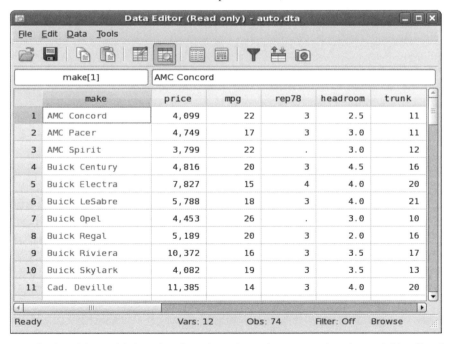

The data are displayed in multiple colors—at first glance it appears that the variables listed in black are numeric, whereas those that are in colors are text. This is worth investigating. Click on a cell under the `make` variable: the input box at the top displays the make of the car. Scroll to the right until you see the `foreign` variable. Click on one of its cells. Although the cell may display "Domestic", the input box displays a 0. This shows that Stata can store categorical data as numbers but display human-readable text. This is done by what Stata calls *value labels*. Finally, under the `rep78` variable, which looks to be numeric, there are some cells containing just a period (`.`). As we will see, these correspond to missing values.

Looking at the data in this fashion, though comfortable, lends little information about the dataset. It would be useful for us to get more details about what the data are and how the data are stored. Close the Data Editor by clicking on its close button.

We can see the structure of the dataset by *describing* its contents. This can be done either by going to **Data > Describe data > Describe data in memory** in the menus and clicking on **OK** or by typing `describe` in the Command window and pressing *Enter*. Regardless of which method you choose, you will get the same result:

```
. describe
Contains data from /usr/local/stata11/ado/base/a/auto.dta
  obs:            74                          1978 Automobile Data
 vars:            12                          13 Apr 2009 17:45
 size:         3,478 (99.9% of memory free)   (_dta has notes)

              storage  display     value
variable name   type   format      label      variable label

make           str18   %-18s                  Make and Model
price          int     %8.0gc                 Price
mpg            int     %8.0g                   Mileage (mpg)
rep78          int     %8.0g                   Repair Record 1978
headroom       float   %6.1f                   Headroom (in.)
trunk          int     %8.0g                   Trunk space (cu. ft.)
weight         int     %8.0gc                  Weight (lbs.)
length         int     %8.0g                   Length (in.)
turn           int     %8.0g                   Turn Circle (ft.)
displacement   int     %8.0g                   Displacement (cu. in.)
gear_ratio     float   %6.2f                   Gear Ratio
foreign        byte    %8.0g       origin      Car type

Sorted by:  foreign
```

If your listing stops short, and you see a blue —more— at the base of the Results window, pressing the Spacebar or clicking on the blue —more— itself will allow the command to be completed.

At the top of the listing, some information is given about the dataset, such as where it is stored on disk, how much memory it occupies, and when the data were last saved. The bold 1978 Automobile Data is the short description that appeared when the dataset was opened and is referred to as a *data label* by Stata. The phrase `_dta has notes` informs us that there are notes attached to the dataset. We can see what notes there are by typing `notes` in the Command window:

```
. notes
_dta:
  1.  from Consumer Reports with permission
```

Here we see a short note about the source of the data.

Looking back at the listing from describe, we can see that Stata keeps track of more than just the raw data. Each variable has the following:

- A *variable name*, which is what you call the variable when communicating with Stata.
- A *storage type*, which is the way in which Stata stores its data. For our purposes, it is enough to know that types beginning with str are *string*, or text, variables, whereas all others are numeric. See [U] **12 Data**.
- A *display format*, which controls how Stata displays the data in tables. See [U] **12.5 Formats: Controlling how data are displayed**.
- A *value label* (possibly). This is the mechanism that allows Stata to store numerical data while displaying text. See [GSU] **9 Labeling data** and [U] **12.6.3 Value labels**.
- A *variable label*, which is what you call the variable when communicating with other people. Stata uses the variable label when making tables, as we will see.

A dataset is far more than simply the data it contains. It is also information that makes the data usable by someone other than the original creator.

Although describing the data says something about the structure of the data, it says little about the data themselves. The data can be summarized by clicking on **Statistics > Summaries, tables, and tests > Summary and descriptive statistics > Summary statistics** and clicking on the **OK** button. You could also type summarize in the Command window and press *Enter*. The result is a table containing summary statistics about all the variables in the dataset:

```
. summarize
    Variable |        Obs        Mean    Std. Dev.       Min        Max
-------------+--------------------------------------------------------
        make |          0
       price |         74    6165.257    2949.496       3291      15906
         mpg |         74     21.2973    5.785503         12         41
       rep78 |         69    3.405797    .9899323          1          5
    headroom |         74    2.993243    .8459948        1.5          5
-------------+--------------------------------------------------------
       trunk |         74    13.75676    4.277404          5         23
      weight |         74    3019.459    777.1936       1760       4840
      length |         74    187.9324    22.26634        142        233
        turn |         74    39.64865    4.399354         31         51
displacement |         74    197.2973    91.83722         79        425
-------------+--------------------------------------------------------
   gear_ratio |        74    3.014865    .4562871       2.19       3.89
     foreign |         74    .2972973    .4601885          0          1
```

From this simple summary, we can learn a bit about the data. First of all, the prices are nothing like today's car prices—of course, these cars are now antiques. We can see that the gas mileages are not particularly good. Automobile aficionados can gain some feel for other characteristics.

There are two other important items here:

- The make variable is listed as having no observations. It really has no numerical observations, because it is a string (text) variable.
- The rep78 variable has 5 fewer observations than the other numerical variables. This implies that rep78 has five missing values.

Although we could use the summarize and describe commands to get a bird's eye view of the dataset, Stata has a command that gives a good in-depth description of the structure, contents, and values of the variables: the codebook command. Either type codebook in the Command window and press *Enter* or navigate the menus to **Data > Describe data > Describe data contents (codebook)** and click on **OK**. We get a large amount of output that is worth investigating. In fact, we get more output than can fit on one screen, as can be seen by the blue —more— at the bottom of the Results window. Press the Spacebar a few times to get all the output to scroll past. (For more about —more—, see **More** in [GSU] **10 Listing data and basic command syntax**.) Look it over to see that much can be learned from this simple command. You can scroll back in the Results window to see earlier results, if need be. We will focus on the output for make, rep78, and foreign.

To start our investigation, we would like to run the codebook command on just one variable, say, make. We can do this via menus or the command line, as usual. To get the codebook output for make via the menus, start by navigating as before, to **Data > Describe data > Describe data contents (codebook)**. When the dialog appears, there are multiple ways to tell Stata to consider only the make variable:

- We could type make into the *Variables* field.
- Clicking on the *Variables* field of the dialog before clicking on the Variables window directs all output from the Variables window to the *Variables* field. Otherwise, output from the Variables window goes to the Command window.

A much easier solution is to type codebook make in the Command window and then press *Enter*. The result is informative:

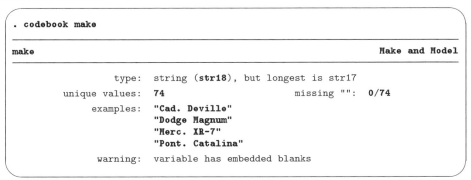

The first line of the output tells us the variable name (make) and the variable label (Make and Model). The variable is stored as a string (which is another way of saying "text") with a maximum length of 18 characters, though a size of only 17 characters would be enough. All the values are unique, so if need be, make could be used as an identifier for the observations—something that is often useful when putting together data from multiple sources or when trying to weed out errors from the dataset. There are no missing values, but there are blanks within the makes. This latter fact could be useful if we were expecting make to be a one-word string variable.

Syntax note: Telling the codebook command to run on the make variable is an example of using a *varlist* in Stata's syntax.

Looking at the foreign variable can teach us about value labels. We would like to look at the codebook output for this variable, and on the basis of our latest experience, it would be easy to type codebook foreign into the Command window to get the following output:

(Continued on next page)

```
. codebook foreign

foreign                                                               Car type

                  type:  numeric (byte)
                 label:  origin
                 range:  [0,1]                      units:  1
         unique values:  2                     missing .:  0/74

            tabulation:  Freq.   Numeric  Label
                            52         0  Domestic
                            22         1  Foreign
```

We can glean that foreign is an indicator variable, because its only values are 0 and 1. The variable has a value label that displays "Domestic" instead of 0 and "Foreign" instead of 1. There are two advantages of storing the data in this form:

- Storing the variable as a byte takes less memory, because each observation uses 1 byte instead of the 8 bytes needed to store "Domestic". This is important in large datasets. See [U] **12.2.2 Numeric storage types**.
- As an indicator variable, it is easy to incorporate into statistical models. See [U] **25 Working with categorical data and factor variables**.

Finally, we can learn a little about a poorly labeled variable with missing values by looking at the rep78 variable. Typing codebook rep78 into the Command window yields

```
. codebook rep78

rep78                                                     Repair Record 1978

                  type:  numeric (int)
                 range:  [1,5]                      units:  1
         unique values:  5                     missing .:  5/74

            tabulation:  Freq.   Value
                            2    1
                            8    2
                           30    3
                           18    4
                           11    5
                            5    .
```

rep78 appears to be a categorical variable, but because of lack of documentation, we do not know what the numbers mean. (To see how we would label the values, see **Changing data** in [GSU] **6 Using the Data Editor** and see [GSU] **9 Labeling data**.) This variable has five missing values, meaning that there are 5 observations for which the repair record is not recorded. We could use the Data Editor to investigate these 5 observations, but we will do this using the Command window only (doing so is much simpler). If you recall from earlier, the command brought up by clicking on the **Data Editor (Browse)** button was browse. We would like to browse only those observations for which rep78 is missing, so we could type

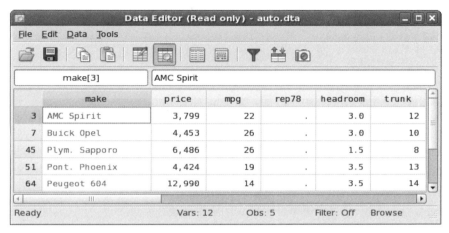

```
. browse if missing(rep78)
```

From this, we see that the . entries are indeed missing values—though other missing values are allowable. See [U] **12.2.1 Missing values**. Close the Data Editor after you are satisfied with this statement.

Syntax note: Using the if qualifier above is what allowed us to look at a subset of the observations.

Looking through the data lends no clues about why these particular data are missing. We decide to check the source of the data to see if the missing values were originally missing or if they were omitted in error. Listing the makes of the cars whose repair records are missing will be all we need, because we saw earlier that the values of make are unique. This can be done via the menus and a dialog:

1. Select **Data > Describe data > List data**.
2. Click within the *Variables* field on the dialog. The title bar for the Variables window changes to *Target: Variables*. This means that clicking on a variable name in the Variables window sends the variable name to the *Variables* field in the dialog.
3. Click on make to enter it into the *Variables* field.
4. Click on the **by/if/in** tab in the dialog.
5. Type missing(rep78) into the *If: (expression)* box.
6. Click on **Submit**. Stata executes the proper command but the dialog box remains open. **Submit** is useful when experimenting, exploring, or building complex commands. We will use primarily **Submit** in the examples. You may click on **OK** in its place if you like.

The same ends could be achieved by typing list make if missing(rep78). The latter is easier, once you know that the command list is used for listing observations. In any case, here is the output:

(Continued on next page)

```
. list make if missing(rep78)

         +-----------------+
         |  make           |
         |-----------------|
   3.    |  AMC Spirit     |
   7.    |  Buick Opel     |
  45.    |  Plym. Sapporo  |
  51.    |  Pont. Phoenix  |
  64.    |  Peugeot 604    |
         +-----------------+
```

We go to the original reference and find that the data were truly missing and cannot be resurrected. See [GSU] **10 Listing data and basic command syntax** for more information about all that can be done with the list command.

Syntax note: This command uses two new concepts for Stata commands—the if qualifier and the missing() function. The if qualifier restricts the observations on which the command runs to only those observations for which the expression is true. See [U] **11.1.3 if exp**. The missing() function tests each observation to see if it is missing. See [U] **13.3 Functions**.

Now that we have a good idea about the underlying dataset, we can investigate the data themselves.

Descriptive statistics

We saw above that the summarize command gave brief summary statistics about all the variables. Suppose now that we became interested in the prices while summarizing the data, because they seemed fantastically low (it was 1978, after all). To get an in-depth look at the price variable, we can use the menus and a dialog:

1. Select **Statistics > Summaries, tables, and tests > Summary and descriptive statistics > Summary statistics**.
2. Enter or select price in the *Variables* field.
3. Select *Display additional statistics*.
4. Click on **Submit**.

Syntax note: As can be seen from the Results window, typing summarize price, detail will get the same result. The portion after the comma contains *options* for Stata commands; hence, detail is an example of an option.

```
. summarize price, detail
                        Price

         Percentiles      Smallest
  1%         3291           3291
  5%         3748           3299
 10%         3895           3667        Obs                    74
 25%         4195           3748        Sum of Wgt.            74

 50%        5006.5                      Mean             6165.257
                           Largest      Std. Dev.        2949.496
 75%         6342          13466
 90%        11385          13594        Variance          8699526
 95%        13466          14500        Skewness         1.653434
 99%        15906          15906        Kurtosis         4.819188
```

From the output, we can see that the median price of the cars in the dataset is only about $5,006! We can also see that the four most expensive cars are all priced between $13,400 and $16,000. If we wished to browse the cars that are the most expensive (and gain some experience with features of the Data Editor), we could start by clicking on the **Data Editor (Browse)** button, . Once the Data Editor is open, we can click the **Filter Observations** button, , to bring up the *Filter Observations* dialog. We can look at the expensive cars by putting `price > 13000` in the *Filter by expression* field:

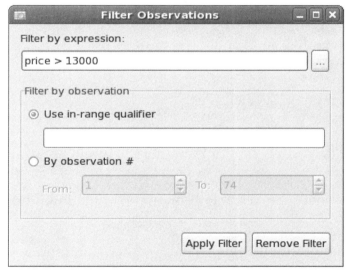

Pressing the **Apply Filter** button filters the data, and we can see that the expensive cars are two Cadillacs and two Lincolns, which have low gas mileage:

(Continued on next page)

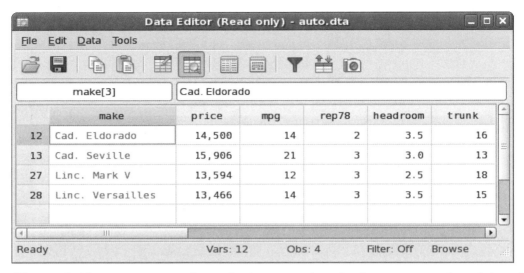

We now decide to turn our attention to foreign cars and repairs, because as we glanced through the data, it appeared that the foreign cars had better repair records. (We do not know exactly what the categories 1, 2, 3, 4, and 5 mean, but we know the Chevy Monza was known for breaking down.) Let's start by looking at the proportion of foreign cars in the dataset along with the proportion of cars with each type of repair record. We can do this with one-way tables. The table for `foreign` cars can be done via menus and a dialog starting with **Statistics > Summaries, tables, and tests > Tables > One-way tables** and then choosing the variable `foreign` in the *Categorical variable* field. Clicking on **Submit** yields

```
. tabulate foreign
   Car type |      Freq.     Percent        Cum.
------------+-----------------------------------
   Domestic |         52       70.27       70.27
    Foreign |         22       29.73      100.00
------------+-----------------------------------
      Total |         74      100.00
```

We see that roughly 70% of the cars in the dataset are domestic, whereas 30% are foreign made. The value labels are used to make the table so that the output is nicely readable.

Syntax note: We also see that this one-way table could be made using the `tabulate` command together with one variable, `foreign`.

Making a one-way table for the repair records is simple—it will be simpler if done via the Command window. Typing `tabulate rep78` yields

```
. tabulate rep78
    Repair |
Record 1978 |      Freq.      Percent        Cum.
-----------+-----------------------------------
         1 |          2         2.90        2.90
         2 |          8        11.59       14.49
         3 |         30        43.48       57.97
         4 |         18        26.09       84.06
         5 |         11        15.94      100.00
-----------+-----------------------------------
     Total |         69       100.00
```

We can see that most cars have repair records of 3 and above, though the lack of value labels makes us unsure what a "3" means. Take our word for it that 1 means a poor repair record and 5 means a good repair record. The five missing values are indirectly evident, because the total number of observations listed is 69 rather than 74.

These two one-way tables do not help us compare the repair records of foreign and domestic cars. A two-way table would help greatly, which we can get by using the menus and a dialog:

1. Select **Statistics > Summaries, tables, and tests > Tables > Two-way tables with measures of association**.
2. Choose `rep78` as the *Row variable*.
3. Choose `foreign` as the *Column variable*.
4. It would be nice to have the percentages within the `foreign` variable, so check the *Within-row relative frequencies* checkbox.
5. Click on **Submit**.

Here is the resulting output:

(Continued on next page)

```
. tabulate rep78 foreign, row
```

Key
frequency
row percentage

Repair Record 1978	Car type Domestic	Foreign	Total
1	2 100.00	0 0.00	2 100.00
2	8 100.00	0 0.00	8 100.00
3	27 90.00	3 10.00	30 100.00
4	9 50.00	9 50.00	18 100.00
5	2 18.18	9 81.82	11 100.00
Total	48 69.57	21 30.43	69 100.00

The output indicates that foreign cars are generally much better than domestic cars when it comes to repairs. If you like, you could repeat the previous dialog and try some of the hypothesis tests available from the dialog. We will abstain.

Syntax note: We see that typing the command `tabulate rep78 foreign, row` would have given us the same table. Thus using `tabulate` with two variables yielded a two-way table. It makes sense that `row` is an option, because we went out of our way to check it in the dialog, so we changed the behavior of the command from its default.

Continuing our exploratory tour of the data, we would like to compare gas mileages between foreign and domestic cars, starting by looking at the summary statistics for each group by itself. We know that a direct way to do this would be to summarize mpg for each of the two values of `foreign`, by using `if` qualifiers:

```
. summarize mpg if foreign==0
```

Variable	Obs	Mean	Std. Dev.	Min	Max
mpg	52	19.82692	4.743297	12	34

```
. summarize mpg if foreign==1
```

Variable	Obs	Mean	Std. Dev.	Min	Max
mpg	22	24.77273	6.611187	14	41

It appears that foreign cars get somewhat better gas mileage—we will test this soon.

Syntax note: We needed to use a double equal sign (==) for testing equality. This could be familiar to you if you have programmed before. If it is unfamiliar, it is a common source of errors when initially using Stata. Thinking of equality as "really equal" can cut down on typing errors.

There are two other methods that we could have used to produce these summary statistics. These methods are worth knowing, because they are less error-prone. The first method duplicates the concept of what we just did by exploiting Stata's ability to run a command on each of a series of nonoverlapping subsets of the dataset. To use the menus and a dialog, do the following:

1. Select **Statistics > Summaries, tables, and tests > Summary and descriptive statistics > Summary statistics** and click on the **Reset** button, **Ⓑ** .

2. Select mpg in the *Variables* field.

3. Select the *Standard display* option (if it is not already selected).

4. Click on the **by/if/in** tab.

5. Check the *Repeat command by groups* checkbox.

6. Select or type foreign for the grouping variable.

7. **Submit** the command.

You can see that the results match those from above. They have a better appearance because the value labels are used rather than the numerical values. The method is more appealing because the results were produced without knowing the possible values of the grouping variable ahead of time.

```
. by foreign, sort: summarize mpg

-> foreign = Domestic
    Variable |      Obs        Mean    Std. Dev.       Min        Max

         mpg |       52    19.82692    4.743297         12         34

-> foreign = Foreign
    Variable |      Obs        Mean    Std. Dev.       Min        Max

         mpg |       22    24.77273    6.611187         14         41
```

There is something different about the equivalent command that appears above: it contains a *prefix command* called a by prefix. The by prefix has its own option, namely, sort, to ensure that like members are adjacent to each other before being summarized. The by prefix command is important for understanding data manipulation and working with subpopulations within Stata. Store this example away, and consult [U] **11.1.2 by varlist:** and [U] **27.2 The by construct** for more information. Stata has other prefix commands for specialized treatment of commands, as explained in [U] **11.1.10 Prefix commands**.

The third method for tabulating the differences in gas mileage across the cars' origins involves thinking about the structure of desired output. We need a one-way table of automobile types (foreign versus domestic) within which we see information about gas mileages. Looking through the menus yields the menu item **Statistics > Summaries, tables, and tests > Tables > One/two-way table of summary statistics**. Selecting this, entering foreign for *Variable 1* and mpg for the *Summarize variable*, and submitting the command yields a nice table:

```
. tabulate foreign, summarize(mpg)
                      Summary of Mileage (mpg)
     Car type            Mean    Std. Dev.          Freq.

     Domestic         19.826923   4.7432972            52
      Foreign         24.772727   6.6111869            22

        Total         21.297297   5.7855032            74
```

The equivalent command is evidently `tabulate foreign, summarize(mpg)`.

Syntax note: This is a one-way table, so `tabulate` uses one variable. The variable being summarized is passed to the `tabulate` command via an option.

A simple hypothesis test

We would like to run a hypothesis test for the difference in the mean gas mileages. Under the menus, **Statistics > Summaries, tables, and tests > Classical tests of hypotheses > Two-group mean-comparison test** leads to the proper dialog. Enter mpg for the *Variable name* and foreign for the *Group variable name*, and **Submit** the dialog. The results are

```
. ttest mpg, by(foreign)
Two-sample t test with equal variances

   Group        Obs        Mean     Std. Err.    Std. Dev.    [95% Conf. Interval]

Domestic         52     19.82692     .657777     4.743297     18.50638    21.14747
 Foreign         22     24.77273     1.40951     6.611187     21.84149    27.70396

combined         74     21.2973      .6725511    5.785503     19.9569     22.63769

    diff               -4.945804    1.362162                  -7.661225   -2.230384

       diff = mean(Domestic) - mean(Foreign)                      t =   -3.6308
Ho: diff = 0                                        degrees of freedom =       72

     Ha: diff < 0                 Ha: diff != 0                    Ha: diff > 0
  Pr(T < t) = 0.0003        Pr(|T| > |t|) = 0.0005            Pr(T > t) = 0.9997
```

From this, we could conclude that the mean gas mileage for foreign cars is different from that of domestic cars (though we really ought to have wanted to test this before snooping through the data). We can also conclude that the command, `ttest mpg, by(foreign)` is easy enough to remember. Feel free to experiment with unequal variances, various approximations to the number of degrees of freedom, and the like.

Syntax note: The `by()` option used here is not the same as the `by` prefix command used earlier. Although it has a similar conceptual meaning, its usage is different because it is a particular option for the `ttest` command.

Descriptive statistics, correlation matrices

We now change our focus from exploring categorical relationships to exploring numerical relationships: we would like to know if there is a correlation between miles per gallon and weight. We select **Statistics > Summaries, tables, and tests > Summary and descriptive statistics > Correlations and covariances** in the menus. Entering mpg and weight, either by clicking or by typing, and then submitting the command yields

```
. correlate mpg weight
(obs=74)

             |      mpg   weight
-------------+------------------
         mpg |   1.0000
      weight |  -0.8072   1.0000
```

The equivalent command for this is natural: correlate mpg weight. There is a negative correlation, which is not surprising, because heavier cars should be harder to push about.

We could see how the correlation compares for foreign and domestic cars by using our knowledge of how the by prefix works. We can reuse the *correlate* dialog or use the menus as before if the dialog is closed. Click on the **by/if/in** tab, check the *Repeat command by groups* checkbox, and enter the foreign variable to define the groups. As done above on page 13, a simple by foreign, sort: prefix in front of our previous command would work, too:

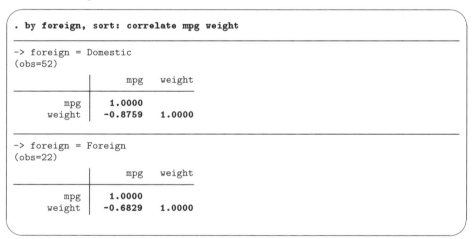

```
. by foreign, sort: correlate mpg weight

-> foreign = Domestic
(obs=52)

             |      mpg   weight
-------------+------------------
         mpg |   1.0000
      weight |  -0.8759   1.0000


-> foreign = Foreign
(obs=22)

             |      mpg   weight
-------------+------------------
         mpg |   1.0000
      weight |  -0.6829   1.0000
```

We see from this that the correlation is not as strong among the foreign cars.

Syntax note: Although we used the correlate command to look at the correlation of two variables, Stata can make correlation matrices for an arbitrary number of variables:

(Continued on next page)

```
. correlate mpg weight length turn displacement
(obs=74)
                       mpg   weight   length     turn displa~t

        mpg       1.0000
     weight      -0.8072   1.0000
     length      -0.7958   0.9460   1.0000
       turn      -0.7192   0.8574   0.8643   1.0000
displacement     -0.7056   0.8949   0.8351   0.7768   1.0000
```

This can be useful, for example, when investigating collinearity among predictor variables. In fact, simply typing `correlate` will yield the complete correlation matrix.

Graphing data

We are about to make some graphs. You must be using Stata(GUI) to see them.

We have found several things in our investigations so far: We know that the average MPG of domestic and foreign cars differs. We have learned that domestic and foreign cars differ in other ways as well, such as in frequency-of-repair record. We found a negative correlation of MPG and weight—as we would expect—but the correlation appears stronger for domestic cars.

We would now like to examine, with an eye toward modeling, the relationship between MPG and weight, starting with a graph. We can start with a scatterplot of `mpg` against `weight`. The command for this is simple: `scatter mpg weight`. Using the menus requires a few steps because the graphs in Stata can be customized heavily.

1. Select **Graphics > Twoway graph (scatter, line, etc.)**.
2. Click on the **Create...** button.
3. Select the *Basic plots* radio button (if it is not already selected).
4. Select *Scatter* as the basic plot type (if it is not already selected).
5. Select `mpg` as the *Y variable* and `weight` as the *X variable*.
6. Click on the **Submit** button.

The Results window shows the command that was issued from the menu:

```
. twoway (scatter mpg weight)
```

The command issued when the dialog was submitted is a bit more complex than the command suggested above. There is good reason for this: the more complex structure allows combining and overlaying graphs, as we will soon see. In any case, the graph that appears is

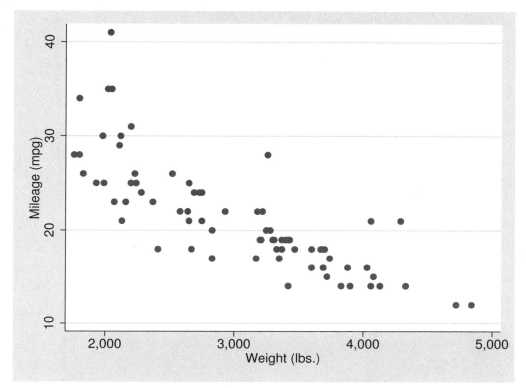

We see the negative correlation in the graph, though the relationship appears to be nonlinear.

Note: When you draw a graph, the Graph window appears, probably covering up your Results window. Click on the main Stata window to get the Results window back on top. Want to see the graph again? Click on the **Graph** button, ▮▮ . See **The Graph button** in [GSU] **14 Graphing data** for more information about the **Graph** button.

Note: You must be using X Windows and Stata(GUI) for the graph to appear. If you are using Stata(console), a graph will not appear when you use the scatter command. For more information on obtaining graphs when using Stata(console), see the *Graphics Reference Manual*.

We would now like to see how the different correlations for foreign and domestic cars are manifested in scatterplots. It would be nice to see a scatterplot for each type of car, along with a scatterplot for all the data.

Syntax note: Because we are looking at subgroups, this looks like it is a job for by. We want one graph—think about whether it should be a by() option or a by prefix.

Start as before:
1. Select **Graphics > Twoway graph (scatter, line, etc.)** from the menus.
2. If *Plot 1* is displayed under the *Plot definitions*, and (scatter mpg weight) appears below the *Plot definitions* box, select it and skip to step 4.
3. Go through the process on the previous page to create the graph.
4. Click on the **By** tab.
5. Check the *Draw subgraphs for unique values of variables* checkbox.
6. Enter foreign in the *Variables* field.

7. Check the *Add a graph with totals* checkbox.

8. Click on the **Submit** button.

The command and the associated graph are

```
. twoway (scatter mpg weight), by(foreign, total)
```

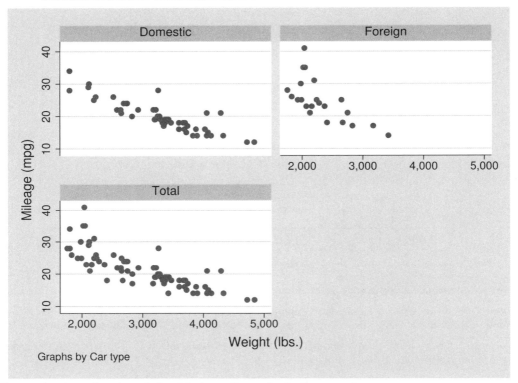

The graphs show that the relationship is nonlinear for both types of cars.

Syntax note: This is another use of a by() option. If you had used a by prefix, two separate graphs would have been generated.

Model fitting: Linear regression

After looking at the graphs, we would like to fit a regression model that predicts MPG from the weight and type of the car. From the graphs, the relationship is nonlinear and so we will try modeling MPG as a quadratic in weight. Also from the graphs, we judge the relationship to be different for domestic and foreign cars. We will include an indicator (dummy) variable for foreign and evaluate afterward whether this adequately describes the difference. Thus we will fit the model

$$\mathtt{mpg} = \beta_0 + \beta_1 \,\mathtt{weight} + \beta_2 \,\mathtt{weight}^2 + \beta_3 \,\mathtt{foreign} + \epsilon$$

foreign is already an indicator (0/1) variable, but we need to create the weight-squared variable. This can be done via the menus, but here using the command line is simpler:

```
. generate wtsq = weight^2
```

Now that we have all the variables we need, we can run a linear regression. We will use the menus and see that the command is also simple. To use the menus, select **Statistics > Linear models and related > Linear regression**. In the resulting dialog, choose mpg as the *Dependent variable* and weight, wtsq, and foreign as the *Independent variables*. **Submit** the command. Here is the equivalent simple regress command and the resulting analysis-of-variance table.

```
. regress mpg weight wtsq foreign

      Source |       SS       df       MS              Number of obs =      74
-------------+------------------------------           F(  3,    70) =   52.25
       Model |  1689.15372     3   563.05124           Prob > F      =  0.0000
    Residual |   754.30574    70  10.7757963           R-squared     =  0.6913
-------------+------------------------------           Adj R-squared =  0.6781
       Total |  2443.45946    73  33.4720474           Root MSE      =  3.2827

         mpg |      Coef.   Std. Err.      t    P>|t|     [95% Conf. Interval]
-------------+----------------------------------------------------------------
      weight |  -.0165729   .0039692    -4.18   0.000    -.0244892   -.0086567
        wtsq |   1.59e-06   6.25e-07     2.55   0.013     3.45e-07    2.84e-06
     foreign |    -2.2035   1.059246    -2.08   0.041      -4.3161   -.0909002
       _cons |   56.53884   6.197383     9.12   0.000     44.17855    68.89913
```

The results look encouraging, so we will plot the predicted values on top of the scatterplots for each of the types of cars. To do this, we need the predicted, or fitted, values. This can be done via the menus, but doing it by hand is simple enough. We will create a new variable, mpghat:

```
. predict mpghat
(option xb assumed; fitted values)
```

The output from this command is simply a notification. Go over to the Variables window and scroll to the bottom to confirm that there is now an mpghat variable. If you were to try this command when mpghat already exists, Stata will refuse to overwrite your data:

```
. predict mpghat
mpghat already defined
r(110);
```

The predict command, when used after a regression, is called a *postestimation command*. As specified, it creates a new variable called mpghat equal to

$$-0.0165729 \, \mathrm{weight} + 1.59 \times 10^{-6} \, \mathrm{wtsq} - 2.2035 \, \mathrm{foreign} + 56.53884$$

For careful model fitting, there are several features available to you after estimation—one is calculating predicted values. Be sure to read [U] **20 Estimation and postestimation commands**.

We can now graph the data and the predicted curve to evaluate the fit on the foreign and domestic data separately to determine if our shift parameter is adequate. We can draw both graphs together. Using the menus and a dialog, do the following:

1. Select **Graphics > Twoway graph (scatter, line, etc.)** from the menus.
2. If there are any plots listed, click on the **Reset** button, ⑧.
3. Create the graph for mpg versus weight.
    a. Click on the **Create...** button.
    b. Be sure that *Basic plots* and *Scatter* are selected.
    c. Select mpg as the *Y variable* and weight as the *X variable*.
    d. Click on **Accept**.
4. Create the graph showing mpghat versus weight.
    a. Click on the **Create...** button.
    b. Select *Basic plots* and *Line*.
    c. Select mpghat as the *Y variable* and weight as the *X variable*.
    d. Check the *Sort on x variable* box. Doing so ensures that the lines connect from smallest to largest weight values, instead of the order in which the data happen to be.
    e. Click on **Accept**.
5. Show two plots, one each for domestic and foreign cars, on the same graph.
    a. Click on the **By** tab.
    b. Check the *Draw subgraphs for unique values of variables* checkbox.
    c. Enter foreign in the *Variables* field.
6. Click on the **Submit** button.

Here are the resulting command and graph:

```
. twoway (scatter mpg weight) (line mpghat weight, sort), by(foreign)
```

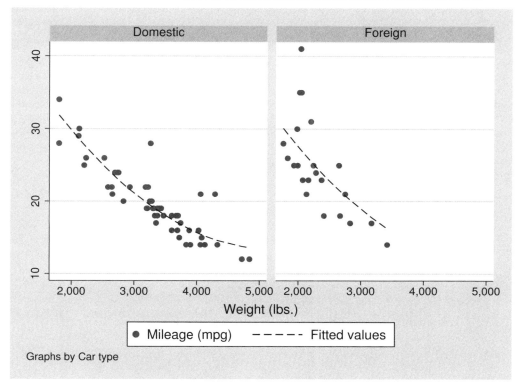

Here we can see the reason for enclosing the separate `scatter` and `line` commands in parentheses: they can then be overlaid by submitting them together. The fit of the plots looks good and is cause for initial excitement. So much excitement, in fact, that we decide to print the graph and show it to an engineering friend. We print the graph, being careful to print the graph and not all our results: **File > Print...** from the Graph window menu bar.

When we show our graph to our engineering friend, she is concerned. "No," she says. "It should take twice as much energy to move 2,000 pounds 1 mile compared with moving 1,000 pounds the same distance, and therefore it should consume twice as much gasoline. Miles per gallon is not a quadratic in weight; gallons per mile is a linear function of weight. Don't you remember any physics?"

We try out what she says. We need to generate a gallons-per-mile variable and make a scatterplot. Here are the commands that we would need—note their similarity to commands issued earlier in the session. There is one new command, the `label variable` command, which allows us to give the gpm variable a *variable label* so that the graph is labeled nicely.

```
. generate gpm = 1/mpg
. label variable gpm "Gallons per Mile"
. twoway (scatter gpm weight), by(foreign, total)
```

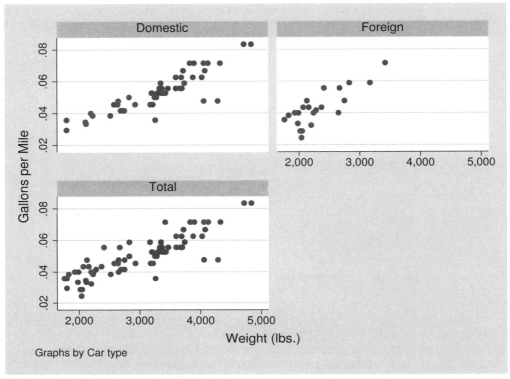

Graphs by Car type

Sadly satisfied that the engineer is indeed correct, we rerun the regression:

```
. regress gpm weight foreign
    Source |      SS          df        MS              Number of obs =      74
-----------+-----------------------------------         F(  2,    71) =  113.97
     Model |  .009117618       2    .004558809          Prob > F      =  0.0000
  Residual |  .00284001       71    .00004              R-squared     =  0.7625
-----------+-----------------------------------         Adj R-squared =  0.7558
     Total |  .011957628      73    .000163803          Root MSE      =  .00632

-----------+-----------------------------------------------------------------
       gpm |      Coef.   Std. Err.      t    P>|t|     [95% Conf. Interval]
-----------+-----------------------------------------------------------------
    weight |   .0000163   1.18e-06    13.74   0.000     .0000139    .0000186
   foreign |   .0062205   .0019974     3.11   0.003     .0022379    .0102032
     _cons |  -.0007348   .0040199    -0.18   0.855    -.0087504    .0072807
-----------------------------------------------------------------------------
```

We find that foreign cars had better gas mileage than domestic cars in 1978 because they were so light. According to our model, a foreign car with the same weight as a domestic car would use an additional 1/160 gallon per mile driven. With this, we are satisfied with our analysis.

Commands versus menus

In this chapter, you have seen that Stata can operate either via menu choices and dialog boxes or via the Command window. As you become more familiar with Stata, you will find that the Command window is typically much faster for often-used commands, whereas the menus and dialogs are faster when building up complex commands, such as graphs.

One of Stata's great strengths is the consistency of its *command syntax*. Most of Stata's commands share the following syntax, where square brackets mean that something is optional and a *varlist* is a list of variables.

$$\left[\,prefix:\,\right]\ command\ \left[\,varlist\,\right]\ \left[\,if\,\right]\ \left[\,in\,\right]\ \left[\,weight\,\right]\ \left[\,,\ options\,\right]$$

Some general rules:

- Most commands accept prefix commands that modify their behavior; see [U] **11.1.10 Prefix commands** for details. One of the more common prefix commands is by.
- If an optional *varlist* is not specified, all the variables are used.
- *if* and *in* restrict the observations on which the command is run.
- *options* modify what the command does.
- Each command's syntax is found in the online help and the reference manuals, or for commands specific to Stata for Unix, in [GSU] **D Stata manual pages for Unix**.
- Stata's command syntax includes more than we have shown you here, but this should get you started. For more information, see [U] **11 Language syntax** and help language.

We saw examples using all the pieces of this except for the in qualifier and the weight clause. The syntax for all commands can be found in the online help along with examples—see [GSU] **4 Getting help** for more information. The consistent syntax makes it straightforward to learn new commands and to read others' commands when examining an analysis.

Here is an example of reading the syntax diagram by using the summarize command from earlier in this chapter. The syntax diagram for summarize is typical:

$$\text{summarize}\ \left[\,varlist\,\right]\ \left[\,if\,\right]\ \left[\,in\,\right]\ \left[\,weight\,\right]\ \left[\,,\ options\,\right]$$

This means that

command by itself is valid:	summarize
command followed by a *varlist* (variable list) is valid:	summarize mpg
	summarize mpg weight
command with *if* (with or without a *varlist*) is valid:	summarize if mpg>20
	summarize mpg weight if mpg>20

and so on.

You can learn about summarize in [R] **summarize**, or select **Help > Stata Command...** and enter summarize, or type help summarize in the Command window.

Keeping track of your work

It would have been useful if we had made a log of what we did so that we could conveniently look back at interesting results or track any changes that were made. You will learn to do this in [GSU] **16 Saving and printing results by using logs**. Your logs will contain commands and their output—another reason to learn command syntax, so that you can remember what you have done.

To make a log file that keeps track of everything appearing in the Results window, click on the button that looks like a lab notebook, . Choose a place to store your log file, and give it a name, just as you would any other document. The log file will save everything that appears in the Results window from the time you start a log file to the time that you close it.

Conclusion

This chapter introduced you to Stata's capabilities. You should now read and work through the rest of this manual. Once you are done here, you can read the *User's Guide*.

2 The Stata user interface

The windows

This chapter introduces the core of Stata (GUI)'s interface: its main windows, its toolbar, its menus, and its dialog boxes. This chapter assumes that you are running Stata(GUI). To reduce clutter, we will refer to the Stata(GUI) as Stata throughout.

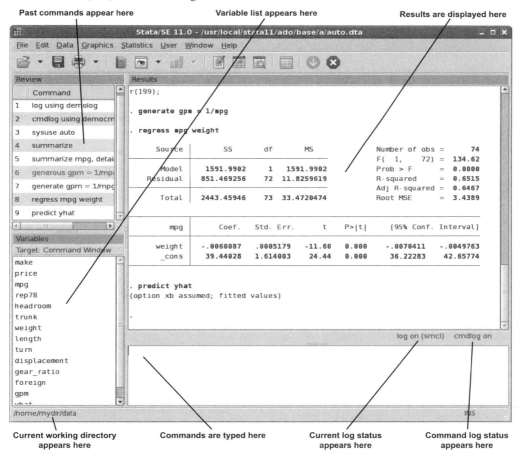

Past commands appear here

Variable list appears here

Results are displayed here

Current working directory appears here

Commands are typed here

Current log status appears here

Command log status appears here

The four main windows are the Review, Variables, Results, and Command windows. Each has its name in its title bar. These four windows are typically in use the whole time Stata is open. There are other, more specialized windows, such as the Viewer, Data Editor, Variables Manager, Do-file Editor, Graph, and Graph Editor windows—these are discussed later in this manual.

To open any window, or to reveal a window hidden by other windows, select the window from the **Window** menu or select the proper item from the toolbar. Many of Stata's windows have functionality that can be accessed by clicking on the right mouse button (right-clicking) within the window. Right-clicking displays a contextual menu that, depending on the window, allows you to copy text, set the preferences for the window, or print the contents of the window. When copying text or printing, we recommend that you always right-click on the window rather than use the menu bar or toolbar so that you can be sure of where and what you are copying or printing.

The toolbar

This is the toolbar:

The toolbar contains buttons that provide quick access to Stata's more commonly used features. If you forget what a button does, hold the mouse pointer over a button for a moment and a tooltip will appear with a description of that button. Buttons that contain both an icon and an arrow display a menu if you click on the arrow.

Open: Open a Stata dataset. Click on the button to open a dataset with the **Open** dialog. Click on the arrow to select a dataset from a menu of recently opened datasets.

Save: Save the Stata dataset currently in memory to disk.

Print Results: Print the Results window. Click on the arrow to select a window to print.

Log: Begin a new log or close, suspend, or resume the current log. See [GSU] **16 Saving and printing results by using logs** for an explanation of log files.

Viewer: Open the Viewer, or bring a Viewer to the front of all other windows. Click on the button to open a new Viewer. Click on the arrow to select a Viewer to bring to the front. See [GSU] **3 Using the Viewer** for more information.

Graph: Bring a Graph window to the front of all other windows. Click on the button to bring the topmost Graph window to the front. Click on the arrow to select a Graph window to bring to the front. See **The Graph button** in [GSU] **14 Graphing data** for more information.

Do-file Editor: Open the Do-file Editor, or bring a Do-file Editor to the front of all other windows. Click on the button to open a new Do-file Editor. Click on the arrow to select a Do-file Editor to bring to the front. See [GSU] **13 Using the Do-file Editor—automating Stata** for more information.

Data Editor (Edit): Open the Data Editor, or bring the Data Editor to the front of the other Stata windows. See [GSU] **6 Using the Data Editor** for more information.

Data Editor (Browse): Open the Data Editor in browse mode. **Browse mode** in [GSU] **6 Using the Data Editor** for more information.

Variables Manager: Open the Variables Manager. See [GSU] **7 Using the Variables Manager** for more information.

Clear —more— Condition: Tell Stata to continue when it has paused in the middle of long output. See [GSU] **10 Listing data and basic command syntax** for more information.

Break: Stop the current task in Stata. See [GSU] **10 Listing data and basic command syntax** for more information.

The Command window

Commands are submitted to Stata from the Command window. The Command window supports basic text editing, copying and pasting, a command history, function-key mapping, and variable-name completion. From the Command window, pressing

> *Page Up*   steps backward through the command history.
> *Page Down*   steps forward through the command history.
> *Tab*   auto-completes a partially typed variable name, if possible.

See [U] **10 Keyboard use** for more information about keyboard shortcuts for the Command window.

The command history allows you to recall a previously submitted command, edit it if you wish, and then resubmit it. Commands submitted by Stata's dialogs are also included in the command history, so you can recall and submit a command without having to open the dialog again.

The Review window

The Review window shows the history of commands that have been entered, displaying successful commands in black and unsuccessful commands, along with their error codes, in red. To enter a command from the Review window, you can

- Click once on a past command to copy it to the Command window, replacing the contents of the Command window.
- Double-click on a past command to execute it. (Executing the command adds the command to the bottom of the Review window, also.)

Right-clicking on the Review window displays a menu from which you can select

- **Cut** to remove the selected command(s) from the Review window and place it on the Clipboard.
- **Copy** to copy the selected command(s) to the Clipboard.
- **Delete** to remove the selected command(s) from the Review window.
- **Select All** to select all the commands in the Review window, including those before and after the commands currently displayed.
- **Clear All** to clear out all the commands from the Review window, including those before and after the commands currently displayed.
- **Do Selected** to submit all the selected commands and add them to the bottom of the command history. Stata will attempt to run all the selected commands, even those containing errors, and will not stop even if a command causes an error.
- **Send to Do-file Editor** to place all the selected commands into a new Do-file Editor window.
- **Save All...** to bring up a *Save Review Contents* dialog, allowing you to save all the commands in the Review window, including those before and after the commands currently displayed, in a *do-file*. (See [GSU] **13 Using the Do-file Editor—automating Stata** for more information on do-files.)
- **Save Selected...** to bring up a *Save Review Contents* dialog, allowing you to save the selected commands in the Review window in a do-file.
- **Preferences...** to edit the preferences for the Review window.

The Variables window

The Variables window shows the list of variables, along with their labels, storage types, and display formats, for the dataset in memory. Click once on a variable in the Variables window to paste it onto the end of whatever is in the Command window. Double-clicking on a variable is no different from clicking on the variable twice—the variable name is entered twice in the Command window. Right-clicking on a variable in the Variables window displays a menu from which you can select

- **Rename** '*varname*'**...** to rename *varname*.
- **Edit Variable Label for** '*varname*'**...** to change the variable label for *varname*.
- **Format** '*varname*'**...** to change the display format for *varname*.
- **Drop** '*varname*'**...** to drop, or eliminate, *varname* from the dataset in memory. You will be asked for confirmation. This affects only the dataset in memory, not the dataset as saved on your disk. See [GSU] **12 Deleting variables and observations** for more information.
- **Manage Notes for** '*varname*'**...** to add, edit, or drop notes attached to variable *varname*. See **Managing notes** in [GSU] **7 Using the Variables Manager** for more information.
- **Manage Notes for Data...** to add, edit, or drop notes attached to the dataset.
- **Preferences...** to change the font used to display the Variables window contents.

Items from the contextual menu issue standard Stata commands, so working by right-clicking is just like working directly in the Command window. You should also investigate the Variables Manager, explained in [GSU] **7 Using the Variables Manager**, because it extends these capabilities and provides a good interface for managing variables.

Menus and dialogs

There are two ways by which you can tell Stata what you would like it to do: you can use menus and dialogs, or you can use the Command window. When you worked through the sample session in [GSU] **1 Introducing Stata—sample session**, you saw that both have strengths. We will discuss the menus and dialogs here.

Stata's **Data**, **Graphics**, and **Statistics** menus provide point-and-click access to almost every command in Stata. As you will learn, Stata is fully programmable, and Stata programmers can even create their own dialogs and menus. The **User** menu provides a place for programmers to add their own menu items. Initially, it contains only some empty submenus.

If you wish to perform a Poisson regression, for example, you could type Stata's `poisson` command, or you could select **Statistics > Count outcomes > Poisson regression**, which would display this dialog:

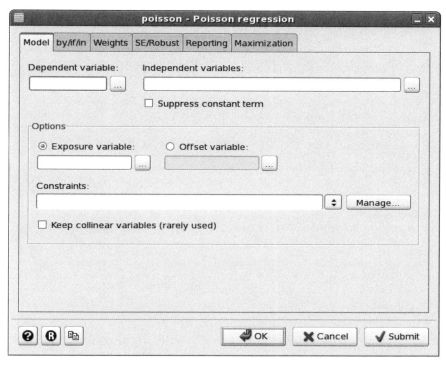

This dialog provides access to all the functionality of Stata's `poisson` command. The `poisson` command has many options that can be accessed by clicking on the multiple tabs across the top of the dialog. The first time you use the dialog for a command, it is a good idea to look at the contents of each tab so that you will know all the dialog's capabilities.

The dialogs for many commands have the **by/if/in** and **Weights** tabs. These provide access to Stata's commands and qualifiers for controlling the estimation sample and dealing with weighted data. See [U] **11 Language syntax** for more information on these features of Stata's language.

The dialogs for most estimation commands have the **Maximization** tab for setting the maximization options (see [R] **maximize**). For example, you can specify the maximum number of iterations for the optimizer.

Most dialogs in Stata provide the same six buttons you see at the bottom of the Poisson dialog above.

 OK issues a Stata command based on how you have filled out the fields in the dialog and then closes the dialog.

 Cancel closes the dialog without doing anything—just as clicking on the red close button does.

Submit issues a command just like **OK** but leaves the dialog on the screen so that you can make changes and issue another command. This feature is handy when, for example, learning a new command or putting together a complicated graph.

 Help provides access to Stata's help system. Clicking on this button will typically take you to the help file for the Stata command associated with the dialog. Clicking on it here would take you to the `poisson` help file. The help file will have tabs above groups of options to show which dialog tab contains which options.

 Reset resets the dialog to its default state. Each time you open a dialog, it will remember how you last filled it out. If you wish to reset its fields to their default values at any time, simply click on this button.

 Copy Command to Clipboard behaves much like the **Submit** button, but rather than issuing a command, it copies the command to the Clipboard. The command can then be pasted elsewhere (such as the Do-file Editor).

The command issued by a dialog is submitted just as if you typed it by hand. You can see it in the Results window and the Review window after it executes. Looking carefully at the full command will help you learn Stata's command syntax.

In addition to being able to access the dialogs for Stata commands through Stata's menus, you can also invoke them by using two other methods. You may know the name of a Stata command for which you want to see a dialog, but you may not remember how to navigate to that command in the menu system. Simply type db *commandname* to launch the dialog for *commandname*:

```
. db poisson
```

You will also find access to the dialog for a command in that command's help file; see [GSU] **4 Getting help** for more details.

As you read this manual, we will present examples of Stata commands. You may type those examples as presented, but you should also experiment with submitting those commands by using their dialogs. Use the db command described above to quickly launch the dialog for any command that you see in this manual.

The working directory

If you look at the screen shot on page 25 you will notice the status bar at the base of the main Stata window that contains /home/mydir/data. This indicates that /home/mydir/data is the current working directory. The current working directory is the folder where graphs and datasets will be saved when typing commands such as save *filename*. It does not affect the behavior of menu-driven file actions such as **File > Save** or **File > Open...**. Once you have started Stata, you can change the current working directory with the cd command. See [D] **cd** for full details. Stata always displays the name of the current working directory so that it is easy to tell where your graphs and datasets will be saved.

3 Using the Viewer

The Viewer in Stata(GUI)

In this chapter, we are going to assume that you are using Stata(GUI) and therefore have access to Stata's menus, links, and windows. If you are not using Stata(GUI), please be aware that you can obtain the same information, but instead of using the menus, links, and windows, you will use commands at the Stata dot prompt. Throughout the following chapters, we will include the equivalent commands that the Stata(console) user can use. The remaining manuals in the Stata Documentation Set also assume that you are using Stata(GUI).

The Viewer's purpose

The Viewer is a versatile tool in Stata(GUI). It will be the first place you can turn for help within Stata, but it is far more than just a help system. You can also use the Viewer to keep your copy of Stata current; add, delete, and manage third-party extensions to Stata known as *user-written programs*; view and print Stata logs both from your current and previous Stata sessions; view and print any other Stata-formatted (SMCL) or plain-text (ASCII) file; and even launch your web browser to follow hyperlinks.

This chapter focuses on the general use of the Viewer, its buttons, and a brief summary of the commands that the Viewer understands. There is more information about using the Viewer to find help in [GSU] **4 Getting help** and for maintaining your copy of Stata in [GSU] **19 Updating and extending Stata—Internet functionality**.

To open a new Viewer window, you can either click on the **Viewer** button, 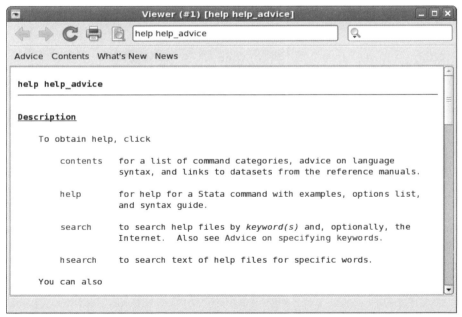, or select **Window > Viewer > New Viewer**. A Viewer opened in this fashion provides links that allow you to perform several tasks.

Viewer buttons

The toolbar of the Viewer has multiple buttons, a command box, and a search box.

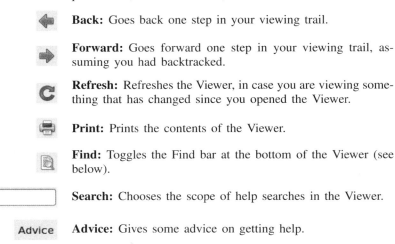

Back: Goes back one step in your viewing trail.

Forward: Goes forward one step in your viewing trail, assuming you had backtracked.

Refresh: Refreshes the Viewer, in case you are viewing something that has changed since you opened the Viewer.

Print: Prints the contents of the Viewer.

Find: Toggles the Find bar at the bottom of the Viewer (see below).

Search: Chooses the scope of help searches in the Viewer.

Advice: Gives some advice on getting help.

Contents: Shows a brief topics table of contents for the help system with links to each major category.

What's New: Lists additions and fixes made to Stata since the initial release of version 11.

News: Lists recent news and information of interest to Stata users.

The Find bar is used to find text within the current Viewer. To reveal the Find bar at the bottom of the window, click on the **Find** button (see above):

✖	Find:		Find Next	Find Previous	☑ Highlight All	☐ Match Case

The Find bar has its own buttons, fields, and checkboxes.

Close: Closes the Find bar.

Find: The field for entering the search text you would like to find.

Find Next: Jumps to the next instance of the search text. Automatically wraps past the end of the Viewer document if there are no further instances of the search text.

Find Previous: Jumps to the previous instance of the search text. Automatically wraps past the start of the Viewer document if there are no previous instances of the search text.

Highlight All: If checked, highlights other instances of the search text (in yellow, by default). If unchecked, only the current instance of the search text is highlighted (in black, by default). By default, this is checked.

Match Case: If checked, uppercase and lowercase letters are considered different, so searching for This would not find this. If unchecked, uppercase and lowercase letters are considered the same, so searching for This would find this. By default, this box is unchecked.

Viewer's function

The Viewer is similar to a web browser. It has clickable links (shown in blue text) that you can follow to related help topics, to install and manage third-party software, and even to keep your copy of Stata up to date. When you move the mouse pointer over a link, the status bar at the bottom of the Viewer shows the action associated with that link. If the action of a link is help logistic, clicking on that link will show the help file for the logistic command in the Viewer. Middle-clicking on a link in a Viewer window (if you do not have a three-button mouse, then *Shift*+clicking will open the link in a new Viewer window.

You can open a new Viewer by selecting **Window > Viewer > New Viewer** or clicking on the Viewer button on the toolbar, or by right-clicking on an existing Viewer window and selecting **Open New Viewer**. Entering a help command from the Command window will also open a new Viewer.

To bring a Viewer to the front of all other Viewers, select **Window > Viewer** and choose a Viewer from the list there. Selecting **Close All Viewers** closes all open Viewer windows.

Viewing local text files, including SMCL files

In addition to viewing built-in Stata help files, you can use the Viewer to view Stata Markup and Control Language (SMCL) files, such as those typically produced when logging your work (see [GSU] **16 Saving and printing results by using logs**) and plain-text files. To open a file and view its contents, simply select **File > View...**, and you will be presented with a dialog:

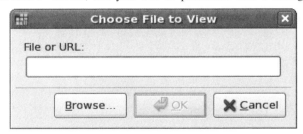

You may either type in the name of the file that you wish to view and click on **OK**, or you may click on the **Browse...** button to open a standard file dialog that allows you to navigate to the file.

If you currently have a log file open, you may view the log file in the Viewer. This method has one advantage over scrolling back in the Results window: what you view stays fixed even as output is added to the Results window. If you wish to view a current log file, select **File > Log > View...**, and the usual dialog will appear, but with the path and filename of the current log already in the field. Simply click on **OK**, and the log will appear in the Viewer. See [GSU] **16 Saving and printing results by using logs** for more details.

Viewing remote files over the Internet

If you want to look at a remote file over the Internet, the process is similar to viewing a local file, only instead of using the **Browse...** button, you type the URL of the file that you want to see, such as http://www.stata.com/man/readme.smcl. You should use the Viewer only to view text or SMCL files. If you enter the URL of, say, an arbitrary web page, you will see the HTML source of the page instead of the usual browser rendering.

Navigating within the Viewer

In addition to using the window scrollbar to navigate the Viewer window, you also can use the up/down cursor keys and *Page Up/Page Down* keys to do the same. Pressing the up/down cursor keys scrolls the window a line at a time. Pressing the *Page Up/Page Down* keys scrolls the window a screenful at a time.

Printing

To print the contents of the Viewer, right-click on the window, and select **Print...**. You may also select **File > Print >** *viewer name* or click on the arrow of the **Print** toolbar button to select from a menu of open windows to print.

Right-clicking on the Viewer window

Right-clicking on the Viewer window displays a contextual menu from which you can do many of the tasks listed above.

- **Back** to go back one step in your viewing trail.
- **Forward** to go forward one step in your viewing trail, assuming you had backtracked.
- **Open New Viewer** to open a new Viewer.
- **Open Link in New Viewer** to open a link in a new Viewer (as long as you right-clicked on the link itself).
- **Close All Viewers But This One** to close all other Viewer windows.
- **Copy** to copy highlighted text to the Clipboard.
- **Find in Viewer...** to find text in the Viewer by using the Find bar.
- **Preferences...** to edit the preferences for the Viewer window.
- **Print...** to print the contents of the Viewer window.

Searching for help in the Viewer

The *search box* in the Viewer can be used to search documentation. Click on the magnifying glass; choose *Search documentation and FAQs*, *Search net resources*, or *Search all*; and then type a word or phrase in the search window and press *Enter*. For more extensive information about using the Viewer for help, see [GSU] **4 Getting help**.

Commands in the Viewer

Everything you can do in the Viewer by clicking on links and buttons can also be done by typing commands in the *command box* at the top of the window or on the Stata command line. Some of the commands that can be issued in the Viewer are

1. *Obtaining help* (see [GSU] **4 Getting help**)

   Type `contents` to view the contents of Stata's help system.

   Type *commandname* to view the help file for a Stata command.

2. *Searching* (see [GSU] **4 Getting help**)

   Type `search` *keyword* to search documentation and FAQs on a topic.

   Type `search` *keyword*`, net` to search net resources on a topic.

   Type `search` *keyword*`, all` to search both of the above.

3. *User-written programs* (see [GSU] **4 Getting help** and [GSU] **19 Updating and extending Stata—Internet functionality**)

   Type `net from http://www.stata.com/` to find and install *Stata Journal*, STB, and user-written programs from the net.

   Type `ado` to review user-written programs you have installed.

   Type `ado uninstall` to uninstall user-written programs you have installed on your computer.

4. *Updating* (see **Official Stata updates** in [GSU] **19 Updating and extending Stata—Internet functionality**)

   Type `update` to check your current Stata version.

   Type `update query` to check for new official Stata update releases.

   Type `update all` to update your Stata.

5. *Viewing files in the Viewer*

   Type `view` *filename*`.smcl` to view SMCL files.

   Type `view` *filename*`.txt` to view ASCII files.

   Type `view` *filename*`.log` to view ASCII log files.

6. *Viewing files in the Results window*

   Type `type` *filename*`.smcl` in the Command window to view SMCL files in the Results window.

   Type `type` *filename*`.txt` in the Command window to view ASCII files in the Results window.

   Type `type` *filename*`.log` in the Command window to view ASCII log files in the Results window.

7. *Launching your browser to view an HTML file*

   Type `browse` *URL* to launch your browser.

8. *Keeping informed*

   Type `news` to see the latest news from http://www.stata.com.

Using the Viewer from the Command window

Stata allows you to print Viewer windows from the Command window. Each Viewer has a unique name (which is displayed in parentheses in the window's title bar) to help identify it from the command line. For example, the name of a Viewer with the window title *Viewer (#1) [help regress]* is **#1**. Its contents can be printed by typing the command `print @Viewer, name(#1)`. Printing in this way does not bring up a print dialog.

Typing `help` *commandname* in the Command window will bring up a new Viewer showing the requested help.

4 Getting help

Online help

Stata's help system provides a wealth of information to help you learn and use Stata. To find out which Stata command will perform the statistical or data-management task you would like to do, you should generally follow these steps:

1. Select **Help > Search...**, choose *Search documentation and FAQs*, and enter the topic or keyword(s). This search will find information about Stata commands, references to articles in the *Stata Journal* or the *Stata Technical Bulletin* (STB), links to FAQs on Stata's web site, and links to selected external web sites.

2. Read through the **Search** results, and click on the appropriate command name link to open its help file.

3. Read the help file for the command you chose.

4. If you want more in-depth help, click on the link from the name of the command to the PDF documentation, read it, then come back to Stata.

5. If the first help file you went to is not what you wanted, either look at the end of the help file for links to take you to related help files or click on the **Back** button to go back to the previous document and go from there to other help files.

6. With the help file open, click on the Command window and enter the command, or click on one of the dialog links at the top right of the help file to open a dialog for the command.

7. If, at any time, you want to begin again with a new **Search**, enter the new search terms in the search box of the Viewer window.

8. If your *Search documentation and FAQs* search returned no results, you can look for *Stata Journal*, STB, and user-written programs on the Internet (materials available via Stata's `net` command) by entering your search term(s) in the search box, clicking on the magnifying glass, and selecting *Search net resources*.

9. If you select *Search documentation and FAQs*, Stata searches Stata's keyword database. If you select *Search net resources*, Stata searches for *Stata Journal*, STB, and user-written programs available for free download on the Internet; see [GSU] **19 Updating and extending Stata— Internet functionality** for more information.

10. You can also select *Search all*, which is equivalent to choosing both *Search documentation and FAQs* and *Search net resources*. This is the equivalent of Stata's `findit` command.

Let's illustrate the **Help** system with an example. You will get the most from the example if you work along at your computer.

Suppose that we have been given a dataset about antique cars, and we need to know what it contains. Though we still have a vague notion of having seen something like this while working through the example session in [GSU] **1 Introducing Stata—sample session**, we do not remember the proper command.

Start by typing `sysuse auto, clear` in the Command window to bring the dataset into memory. (See [GSU] **5 Opening and saving Stata datasets** for information on the `clear` option.)

Following the above approach, we

1. Select **Help > Search...**.
2. Check that the *Search documentation and FAQs* radio button is selected.

37

3. Enter dataset contents into the search box, and click on **OK** or press *Enter*. Before pressing *Enter*, the window should look like

4. Stata will now search for "dataset contents" among the Stata commands, the reference manuals, the *User's Guide*, the *Stata Journal*, the *Stata Technical Bulletin*, and the FAQs on Stata's web site. Here is the result:

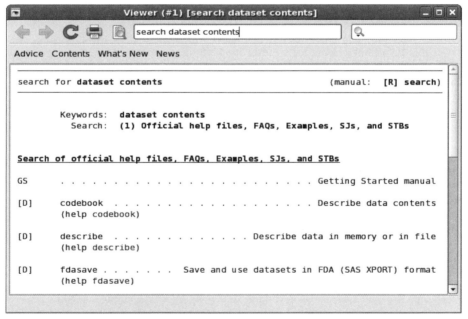

5. Upon seeing the results of the search, we see two commands that look promising: codebook and describe. Because we are interested in the contents of the dataset, we decide to check out the codebook command. The [D] means that we could look up the codebook command in the *Data-Management Reference Manual*. The blue codebook link in (help codebook) means that there is an online help file for the codebook command. This is what we are interested in right now.

6. Click on the blue codebook link. Links can go to a variety of resources, such as help for Stata commands, dialog boxes, and even web pages. Here the link goes to the help file for the codebook command.

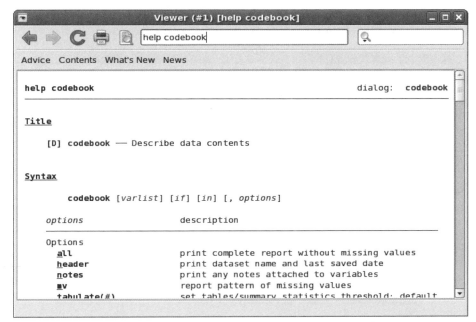

7. What is displayed is typical for help for a Stata command. Help files for Stata commands contain, from top to bottom,

    a. Links to dialog(s) that can be useful in issuing the command. In this example, there is one link to the codebook dialog.

    b. A link to the manual entry for codebook in the PDF documentation. Clicking the link will open your PDF viewer and show you the complete documentation for codebook.

    c. The command's syntax, i.e., rules for constructing a command that Stata will correctly interpret. The square brackets here indicate that all the arguments to codebook are optional but that if we wanted to specify them, we could use a *varlist*, an if qualifier, or an in qualifier, along with some options. (Options vary greatly from command to command.) The options are listed directly under the command and are explained in some detail later in the help file. You will learn more about command syntax in [GSU] **10 Listing data and basic command syntax**.

    d. The location in the menu system for the codebook command.

    e. A description of the command. Because "codebook" is the name for big binders containing hard copy describing each of the elements of a dataset, the description for the codebook command is justifiably terse.

    f. The options that can be used with this command. These are explained in much greater detail than in the listing of the possible options after the syntax. Here, for example, we can see that the mv option can look to see if there is a pattern in the missing values—something important for data cleaning and imputation.

    g. Examples of command usage. The codebook examples are real examples that step through using the command on a dataset either shipped with Stata or loadable within Stata from the Internet.

    h. The information codebook saves in the returned results. These results are used primarily by programmers.

    i. References to related commands, some of which are links to manual entries, some of which are links to help files.

For now, scroll down to the examples. It is worth going through the examples as given in the help file. Here is a screen shot of the top of the examples:

Searching help

Search is designed to help you find information about statistics, graphics, data management, and programming features in Stata. When entering topics for the search, use appropriate terms from statistics, etc. For example, you could enter Mann-Whitney. Multiple topic words are allowed, e.g., regression residuals. For advice on using **Search**, click on the **Advice** button, and follow your choice of advice links.

When you are using **Search**, use proper English and proper statistical terminology. If you already know the name of the Stata command and want to go directly to its help file, select **Help > Stata Command...** and type the command name. You can also type the command name in the *search* field at the top of the Viewer and press *Enter*.

Help distinguishes between topics and Stata commands, because some names of Stata commands are also general topic names. For example, logistic is a Stata command. If you choose **Stata Command...** and enter logistic, you will go right to the help file for the command. But if you choose **Search...** and enter logistic, you will get search results listing the many Stata commands that relate to logistic regression.

Remember that you can search for help from within a Viewer window by typing a command in the Command box of the Viewer or by selecting the scope of the search by using the button to the left of the Search box, typing the search criteria in the Search box, and pressing *Enter*.

Contents

If you click on the **Contents** button, you will see several help categories.

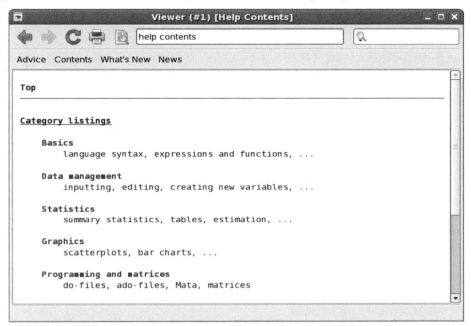

Click on a link to display more information about that category. For example, if you click on **Statistics**, more information about statistics will be displayed.

Help and search commands

As you may expect, the help system is accessible from the Command window. This feature is especially convenient when you need help on a particular Stata command. Here is a short listing of the various commands you can use:

- Typing search *topic* in the Command window produces the same output as selecting **Help > Search...**, choosing *Search documentation and FAQs*, and entering *topic*. The output appears in the Results window, not in a Viewer.
- Typing search *topic*, net in the Command window produces the same output as selecting **Help > Search...**, choosing *Search net resources*, and entering *topic*. The output appears in the Results window, not in a Viewer.
- Typing search *topic*, all in the Command window produces the same output as selecting **Help > Search...**, choosing *Search all*, and entering *topic*. The output appears in the Results window, not in a Viewer.
- Typing findit *topic* in the Command window behaves like search *topic*, all, except the output appears in a Viewer window. This is often the fastest way to search.
- Typing help *commandname* is equivalent to selecting **Help > Stata Command...** and entering *commandname*. The help file for the command appears in a new Viewer.
- Typing chelp *commandname* is similar to selecting **Help > Stata Command...** and entering *commandname*, except the help file for the command appears in the Results window instead of a Viewer.

See [U] **4 Stata's help and search facilities** and [U] **4.9 search: All the details** in the *User's Guide* for more information about these command-language versions of the **Help** system. The `search` command, in particular, has a few capabilities (such as author searches) that we have not demonstrated here.

The Stata reference manuals and User's Guide

Notations such as [R] **ci**, [R] **regress**, and [R] **ttest** in the **Search** results and help files are references to the three-volume *Base Reference Manual*. You may also see things like [P] **#delimit**, which is a reference to the *Programming Reference Manual*, and [U] **9 The Break key**, which is a reference to the *User's Guide*. For a complete list of manuals and their shorthand notations, see *Cross-referencing the documentation*, which immediately follows the table of contents in this manual. Many of the links in the help files point to the PDF versions of the manuals that came with Stata. It is worth clicking on these links to read the extensive information found in the manuals. The Stata help system, though extensive, contains only a fraction of the information found in the manuals.

The Stata reference manuals are each arranged like an encyclopedia, alphabetically, and each has its own index. The *User's Guide* also has an index. The *Quick Reference and Index* contains a combined index for the *User's Guide* and all the reference manuals. This combined index is a good place to start when you are looking for information about a command.

Entries have names like **collapse**, **egen**, . . . , **summarize**, which are generally themselves Stata commands.

For advice on how to use the reference manuals, see [GSU] **18 Learning more about Stata**, or see [U] **1.1 Getting Started with Stata**.

The Stata Journal and the Stata Technical Bulletin

The *Search documentation and FAQs* facility searches all the Stata source materials, including the online help, the *User's Guide*, the reference manuals, this manual, the *Stata Journal*, the *Stata Technical Bulletin* (STB), and the FAQs on Stata's web site.

The *Search net resources* facility searches all materials available via Stata's `net` command; see [R] **net**. These materials include the *Stata Journal*, STB, and user-written additions to Stata available on the Internet.

The *Stata Journal* is a printed and electronic journal, published quarterly, containing articles about statistics, data analysis, teaching methods, and effective use of Stata's language. The *Journal* publishes reviewed papers together with shorter notes and comments, regular columns, tips, book reviews, and other material of interest to researchers applying statistics in a variety of disciplines. The *Journal* is a publication for all Stata users, both novice and experienced, with different levels of expertise in statistics, research design, data management, graphics, reporting of results, and of Stata, in particular. See http://www.stata-journal.com for more information. There, you may browse the archive, subscribe, order PDF copies of individual articles, and view at no charge copies of articles older than 3 years.

The predecessor to the *Stata Journal* was the *Stata Technical Bulletin* (STB). Even though the STB is no longer published, past issues contain articles and programs that may interest you. See http://www.stata.com/bookstore/stbj.html for the table of contents of past issues, and see the STB FAQs at http://www.stata.com/support/faqs/res/stb.html for detailed information on the STB.

Associated with each issue of both the *Stata Journal* and the STB are the programs and datasets described therein. These programs and datasets are made available for download and installation over the Internet, not only to subscribers, but to all Stata users. See [R] **net** and [R] **sj** for more information.

Because the *Stata Journal* and the STB have had several articles dealing with regression models, if you select **Help > Search...** and choose *Search documentation and FAQs*, and enter `ordered regression`, you will see some of these references:

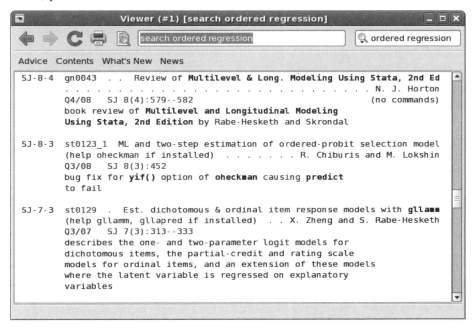

SJ-8-4 refers to volume 8, number 4 of the *Stata Journal*. This is an example of a review article.

SJ-8-3 refers to volume 8, number 3 of the *Stata Journal*. This is an example of a *Stata Journal* article that updates the `oheckman` user-written command. The command can be downloaded and installed to extend the abilities of Stata. See **Downloading user-written programs** in [GSU] **19 Updating and extending Stata—Internet functionality** for more information.

Clicking the SJ links will open a browser and take you to the Stata Press web site, where you can download abstracts and articles. The Stata Press web site allows all articles that are older than three years old to be downloaded for free.

Links to other sites where you can freely download programs and datasets for Stata can be found on the Stata web site; see http://www.stata.com/links/. See **Downloading user-written programs** in [GSU] **19 Updating and extending Stata—Internet functionality** for more details on how to install this software. Also see [R] **ssc** for information on a convenient interface to resources available from the Statistical Software Components (SSC) archive.

We recommend that all users subscribe to the *Stata Journal*. See [U] **3.5 The Stata Journal** for more information.

Notes

5 Opening and saving Stata datasets

How to load your dataset from disk and save it to disk

Opening and saving datasets in Stata works similarly to those tasks in other computer applications. There are a few differences, however. First off, it is possible to save and open files from within Stata's Command window. Second, Stata allows just one dataset to be open and in use at any one time. It is possible to have many Viewers viewing many files, but only one dataset may be in use at any time. Keeping this in mind will make Stata's care in opening new datasets clear. This chapter outlines all the possible ways to open and save datasets.

A Stata dataset can be opened in a variety of ways, most of which are probably familiar to you from other applications:

- Double-click on a Stata data file, which is a file whose extension is dta. Note: The file extension may not be visible, depending on what options you have set in your operating system.
- Select **File > Open...** or click on the **Open** button and navigate to the file.
- Select **File > Open Recent >** *filename*.
- Type use *filename* in the Command window. Stata will look for *filename* in the current working directory. If the file is located elsewhere, you will need to give its path.

Because Stata has at most one dataset open at a time, opening a dataset will cause Stata to discard the dataset that is currently in memory. If there have been changes to the data in the currently open dataset, Stata will refuse to discard the dataset unless you force it to do so. If you open the file with any method other than the Command window, you will be prompted. If you use the Command window and the current data have changed, you will get the following error:

```
. sysuse auto
no; data in memory would be lost
r(4);
```

These behaviors protect you from mistakenly losing data.

To save an unnamed dataset (or an old dataset under a new name):
1. Select **File > Save As...**.
2. Type save *filename* in the Command window.

To save a dataset for use with Stata 8 or Stata 9:
1. Select **File > Save As...**, and select **Stata 9 Data (*.dta)** from the list of file types.
2. Type saveold *filename* in the Command window.

To save a dataset that has been changed (overwriting the original data file):
3. Simply type save, replace in the Command window.

Once you overwrite a dataset, there is no way to recover your original dataset. With important datasets, you may want to either keep a backup copy of your original *filename*.dta or save your changes to a dataset under a new name. This is no different from working with a word-processing document, except that recovering from an inadvertent save with a dataset is nearly impossible.

Important note: Changes you have made to a dataset are not permanent until you save them. You work with a copy of the dataset in memory, not with the data file itself. This should not be surprising, because it is the way that you work with most all applications on your computer.

If you do not want to save your dataset, you can clear the dataset in memory and open a new dataset by typing use *filename*, `clear`.

Troubleshooting

If you try to open a dataset and Stata complains that there is not enough memory, your dataset is larger than the amount of memory that Stata is using. This can happen because Stata must load the entire dataset into memory before it can be used. (By default, Stata/MP and Stata/SE use 50 MB of memory for data, and Stata/IC uses 10 MB of memory.) Changing the amount of memory Stata uses can be done easily from within Stata. For full details, see [GSU] **B Managing memory**.

6 Using the Data Editor

The Data Editor in Stata(GUI)

This chapter discusses the Data Editor for Stata(GUI). Stata(console) users should consult [D] **input** to learn how to input data interactively.

The Data Editor gives a spreadsheet-like view on data, if any, that are currently in memory. You can use it to enter new data, edit existing data, and edit attributes of the data in the dataset, such as variable names, labels, and display formats, as well as value labels.

Any action you take in the Data Editor results in a command being issued to Stata as though you had typed it into the Command window. This means that you can keep good records and learn commands by using the Data Editor.

The Data Editor can be kept open while you work in Stata, giving you a live view of your dataset as you work. To protect your data from inadvertent changes, the Data Editor has two modes: edit mode for active editing and browse mode for viewing. In browse mode, editing within the Data Editor window is disabled. We highly recommend that you use the Data Editor in browse mode by default and switch to edit mode only when you want to make changes.

We will be entering and editing data in this chapter, so start the Data Editor in edit mode by clicking the **Data Editor (Edit)** button, .

Note: Stata(console) users, see [D] **input** to learn how to input data interactively.

Buttons on the Data Editor

The toolbar for the Data Editor has some standard buttons and some buttons we have not yet seen:

Open: Open a Stata dataset. Stata will warn you if your current dataset has unsaved changes.

Save: Save the dataset visible in the Data Editor.

Copy: Copies the current selection to the Clipboard.

Paste: Paste the contents of the Clipboard. You may paste only if one cell is selected—this cell will become the upper left corner of the pasted contents. Warning: This will paste over existing data.

Data Editor (Edit): Changes the Editor to edit mode.

Data Editor (Browse): Changes the Editor to browse mode for safely looking at data.

Variables Manager: Opens the Variables Manager. See [GSU] **7 Using the Variables Manager** for more information.

Variable Properties: Opens a dialog for editing the properties of the highlighted variable.

 Filter Observations: Filters the observations visible in the Data Editor. Useful for looking at a subset of the current dataset.

 Hide/Show Variables: Opens the *Hide/Show Variables* window. Allows viewing a subset of the variables.

 Snapshots: Opens the *Snapshots* window. See **Working with snapshots** below.

You can move about in the Editor with typical methods:
- To move to the right, use the *Tab* key or the right arrow key.
- To move to the left, use *Shift+Tab* or the left arrow key.
- To move down, use *Enter* or the down arrow key.
- To move up, use *Shift+Enter* or the up arrow key.

You can also click within a cell to select it.

Right-clicking within the Data Editor brings up a contextual menu that allows you to manipulate the data as well as what you are viewing. We will cover much of this below. Right-clicking on the Data Editor window displays a menu from which you can do many common tasks:
- **Copy** to copy data to the Clipboard.
- **Paste** to paste data from the Clipboard.
- **Select All** to select all the data displayed in the Data Editor. This could be different from the data in the dataset if the data are filtered or some variables are hidden.
- **Variable Properties...** to display the *Variable Properties* window for the selected variable(s).
- **Replace Contents of Variable...** to bring up a dialog for replacing the values of the selected variable.
- **Sort...** to sort the dataset by the selected variable.
- **Hide Selected Variables** to hide the selected variable(s).
- **Show Only Selected Variables** to hide all but the selected variable(s).
- **Show Entire Dataset** to turn off all filters and unhide all variables.
- **Keep Only Selected Data** to keep only the selected data in the dataset. All remaining data will be dropped (removed) from the dataset. As always, this affects only the data in memory. It will not affect any data on disk.
- **Drop Selected Data** to drop the selected data. This is only possible if the selected data consist of either specific variables or specific observations.
- **Value Labels** to access a submenu for managing and displaying value labels.
- **Preferences...** to set the preferences for the Data Editor.

Data entry

Entering data into the Data Editor is similar to entering data into a spreadsheet. One difference is that the Data Editor has the concept of observations, which makes the data entry smart. We will illustrate this with an example. It will be useful for you to follow the example at your computer. You will need to start with an empty dataset to work along, so save your dataset if necessary, and then type `clear` in the Command window.

Note: As a check to see if your data have changed, type `describe, short` (or `d,s` for short). Stata will tell you if your data have changed.

Suppose that we have the following dataset:

Make	Price	MPG	Weight	Gear Ratio
VW Rabbit	4697	25	1930	3.78
Olds 98	8814	21	4060	2.41
Chev. Monza	3667		2750	2.73
AMC Concord	4099	22	2930	3.58
Datsun 510	5079	24	2280	3.54
	5189	20	3280	2.93
Datsun 810	8129	21	2750	3.55

We do not know MPG for the third car or the make of the sixth.

Start by opening the Data Editor in edit mode. You can do this either by clicking on the **Data Editor (Edit)** button, ![button], or by typing edit in the Command window. You should be greeted by a Data Editor with no data displayed. (If you see data, type clear in the Command window.) Stata shows the active cell by highlighting it and displaying *varname[obsnum]* next to the input box in the *Cursor Location* box. We will see below that we can navigate within a dataset by using this cell reference. The Data Editor starts, by default, in the first row of the first column. Because there are no data, there are no variable names, and so Stata shows var1[1] as the active cell.

We can enter these data either by working across the rows (observation by observation) or by working down the columns (variable by variable). To enter the data observation by observation, press *Tab* after entering each value until you have reached the end of the first row. In our case, we would type VW Rabbit, press *Tab*, type 4697, press *Tab*, and continue entering data to complete the first observation.

After you are finished with the first observation, select the second cell in the first column either by clicking within it or by navigating to it. At this point, your screen should look like this:

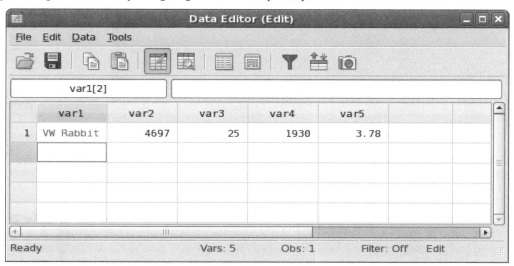

We can now enter the data for the second observation in the same fashion as the first—with one nice difference: after entering the last value in the row, pressing the *Tab* key will bring us to the first cell in the third row. This is possible because the number of variables is known after the first observation has been entered, so Stata knows when it has all the data for an observation.

We can enter the rest of the data by pressing the *Tab* key between entries, simply skipping over missing values by tabbing through them.

If we had wanted to enter the data variable by variable, we could have done that by pressing *Enter* between each make of car until all 7 observations were entered, skipping past the missing entry by using *Enter* twice in succession. Once the first variable was entered we would select the first cell in the second column and enter the price data. We would continue this until we were done.

Notes on data entry

There are several things to note about data entry and the feedback you get from the Data Editor as you enter data:

- *Stata does not allow blank columns or rows in the middle of your dataset.*
  Whenever you enter new variables or observations, always begin in the first empty column or row. If you skip columns or rows, Stata will fill in the intervening columns or rows with missing values.

- *Strings and value labels are color coded.*
  To help distinguish between the different types of variables in the Data Editor, string values are displayed in red, value labels (see [GSU] **9 Labeling data**) are displayed in blue, and all other values are displayed in black. You can change the colors for strings and value labels by right-clicking on the Data Editor window and selecting **Preferences...**.

- *A period ('.') represents Stata's system missing numeric value.*
  Stata has a system missing value, '.', and extended missing values '.a' through '.z'. By default, Stata uses its system missing value.

- *The Tab key is smart.*
  As we saw above, after the first observation has been entered, Stata knows how many variables you have. So at the end of the second observation (and all subsequent observations), *Tab* will automatically take you back to the first column.

- *The Cursor Location box both shows location and is used for navigation.*
  The Cursor Location box gives the location of the current cell. If you see, for example, var3[4], this means that the current cell is the fourth observation of the variable named var3. You can navigate to a particular cell by typing the variable name and the observation in the Cursor Location box. If you wanted the second observation of var1 to be the active cell, typing var1 2 in the Cursor Location box and pressing *Enter* would take you there.

- *Quotes around text are unnecessary in string variables.*
  Once Stata knows that a variable is a string variable (it holds text), there is no need to put quotes around the values, even if the values look like a number. Thus, if you wanted to enter ZIP codes as text, you would enter the first ZIP code with quotes ("02173"), but the rest would not need any quotes.

- If you select a cell and type new data, using an arrow key will accept the change and move to a new active cell. If you double-click on a cell, you can edit within the cell contents. In this case, the arrow keys move within the cell's data.

- If, while you are entering data in a cell, you decide you would like to cancel the changes, press the *Esc* key or click outside the cell.

Renaming and formatting variables

The data have now been entered into Stata, but the variable names leave something to be desired: they have the default names `var1`, `var2`, ..., `var5`. We would like to rename the variables so that they match the column titles from our dataset. We would also like to give the variables descriptions and change their formatting.

We will step through changing the name, format, and label of the `price` variable. Start by right-clicking somewhere within the column for `price` and select **Variable Properties...** from the contextual menu. This opens the *Variable Properties* window where we can change the name, variable label, display format, and value label. Once the dialog box is open,

1. Double-click within the *Name* field to select the old variable name, `var2`, and type `price` to overwrite the name.
2. Press *Tab* to change focus to the *Label* field.
3. Enter a worthwhile label, such as `Price in Dollars`.
4. Click the **...** button next to the *Format* field. You can see here that there are many possible formats, most of which are related to time. We want commas in our numbers, so check the *Use commas in numeric output* checkbox. When you are done, click the **OK** button.
5. These are all the changes for `price`, so click the **Apply** button to make the changes stick.

To edit the properties of another variable, either click within that variable's column or use the navigation arrows in the *Variable Properties* window until the variable's name appears. We can name the first variable `make`, the third `mpg`, the fourth `weight`, and the fifth `gear_ratio`. Just before renaming `var5` to `gear_ratio`, your screen should look like this:

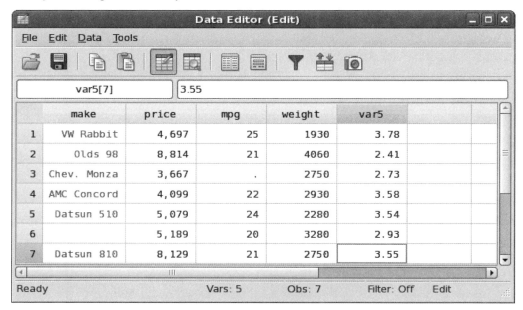

You need to know some rules for variable names:

- *Stata is case sensitive.*
  Make, make, and MAKE are all different names to Stata. If you had named your variables Make, Price, MPG, etc., then you would have to type them correctly capitalized in the future. Using all lowercase letters is easier.
- *A variable name must be 1–32 characters long.*

- *The characters can be letters* (A–Z, a–z), *digits* (0–9), *or underscores* (_).
- *Spaces or other characters are not allowed.*
- *The first character of a variable name must be a letter or an underscore.*
  Although you can use an underscore to begin a variable name, it is highly discouraged. Such names are used for temporary variable names in Stata. You would like your data to be permanent, so using a temporary name could lead to great frustration.

For more information about variable and value labels, see [GSU] **9 Labeling data**; for display formats, see [U] **12.5 Formats: Controlling how data are displayed**.

Copying and pasting data

You can copy and paste data by using the Data Editor. This is often a simple way to bring data into Stata from any other applications such as spreadsheets or databases.

1. *Select the data that you wish to copy.*
   a. Click once on a variable name or column heading to select an entire column.
   b. Click once on an observation number or row heading to select the entire row.
   c. Click and drag the mouse to select a range of cells.
2. *Copy the data to the Clipboard.*
   Right-click within the selected range, and select **Copy**.
3. *Paste the data from the Clipboard.*
   a. Click on the top left cell of the area to which you wish to paste.
   b. Right-click on the same cell, and select **Paste**.

We will illustrate copying and pasting an observation by making a copy of the first observation and pasting it at the end of the dataset.

Start by clicking on the observation number of the first observation. Doing so highlights all the data in the row. Right-click on the same location (there is no need to move the mouse), and select **Copy**:

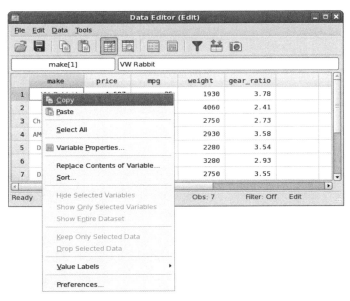

Click on the first cell in the eighth row, right-click while you are still in that cell, and choose **Paste** from the resulting menu. You can see that the observation was successfully duplicated.

Notes on copying and pasting

- The above example illustrated copying and pasting within the Data Editor. You can use roughly the same technique to copy and paste between other applications and Stata, and between Stata and other applications. The one requirement for things to work well is that the external application must copy tables in either tab-delimited or comma-delimited form, as do spreadsheet applications, many database applications, and some word processors. The easiest way to see if copying and pasting works properly is to try it. For more information on file-based methods for importing data into Stata, see [GSU] **8 Importing data**.

- If you are copying and pasting data with value labels, you have a choice. You can copy variables with value labels as text, using the value labels as the actual values, or you can copy said variables as their underlying encoded numbers. Copying with the value labels is the default. If you would like the other choice, select **Tools > Value Labels > Hide All Value Labels**.

Changing data

As its name suggests, the Data Editor can be used to edit your dataset. As we have seen already, it can be used to edit the data themselves as well as the description and display options for the variables.

Here is an example for making some changes to the `auto` dataset, which illustrates both methods for using the Data Editor and its documentation trail. We will also keep *snapshots* of the dataset as we are working, so that we can revert to previous versions of the dataset in case we make a mistake.

We would like to investigate the dataset, work with value labels, delete the `trunk` variable, and make a new variable showing gas consumption per 100 miles. These tasks will illustrate the basics of working in the Data Editor.

Start by typing `sysuse auto` into the Command window. You get an error and are told that your data have changed. This is good—Stata is keeping you from inadvertently destroying your data. If you would like to save the dataset, select **File > Save** and save the dataset in an appropriate location. Otherwise, type `clear` in the Command window and press *Enter* to clear out the data.

Once the `auto` dataset is loaded, start the Data Editor.

1. We would like to see which cars have the lowest and highest gas mileages. To do this, right-click within the `mpg` column. Select **Sort...** from the contextual menu. A dialog will pop up asking if you want to sort. Click on **Yes**. (Stata worries about sort order, because sort order can affect reproducibility when using resampling techniques. This is a good thing.) You will see that the data have now been sorted by `mpg` in ascending order. The lowest-mileage cars are at the top of the screen; by scrolling to the bottom of the dataset, you can find the highest-mileage cars. You also could have sorted by selecting **Tools > Sort...** once the `mpg` variable was selected. For more general sorting, you could use **Data > Sort**, and pick what you need.

2. We would like to investigate repair records, and hence sort by the `rep78` variable. (Do this now.) We see that the Starfire and Firebird both had poor repair records, but we would like to see the cars with good repair records. We could scroll to the bottom of the dataset, but it will be faster to use the *Cursor Location* box: type `rep78 74` and press *Enter* to make `rep78[74]` the active cell. We notice that the last 5 entries for `rep78` are missing. A few items of note:

   a. As we can see from the result of the sort, Stata views missing values as being larger than all numeric nonmissing values: `rep78 >= .` is equivalent to `missing(rep78)`.

    b. What we do not see here is that Stata has multiple missing-value indicators: '.' is Stata's default or system missing-value indicator, and .a, .b, ..., .z are Stata's extended missing values. Extended missing values are useful for indicating the reason why a value is unknown.

    c. The different missing values sort within themselves: . < .a < .b < ··· < .z. See [U] **12.2.1 Missing values** for full details.

3. We would like to make the repair records readable. To start this procedure, right-click within any cell—preferably on a cell in the rep78 column (though this is not necessary). From the contextual menu, choose **Value Labels > Manage Value Labels...**. We need to define a new value label for the repair records.

    a. Click the **Create Label** button. You will see the *Create Label* dialog.

    b. Type a name for the label, say, repairs, in the *Label name* box.

    c. Press the *Tab* key or click within the *Value* field.

    d. Type 1 for the value, press the *Tab* key, and enter atrocious for the label.

    e. Press the *Enter* key to create the pairing.

    f. Repeat steps d and e to make all the pairings: 2 with "bad", 3 with "OK", 4 with "good", and 5 with "stupendous".

    g. Click the **OK** button to finish creating the value.

    h. Click the expose button, ▷ , to show the label—you should see this:

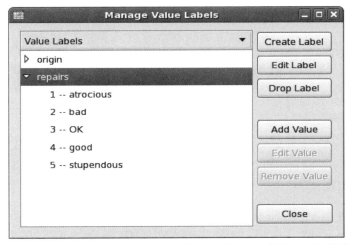

If you have something else, you can edit the label by clicking the **Edit Label** button.

    i. Click on the **Close** button to close the *Manage Value Labels* dialog.

4. Now that the label has been created, attach it to the rep78 variable by right-clicking within any cell in the rep78 column and selecting the **Value Labels > Assign Value Label to Variable 'rep78' > repairs** menu item. You can see that the labels now display in place of the values.

5. Suppose that we found the original source of the data in a time capsule, so we could replace some of the missing values for rep78. We could type the values into cells. We can also assign the values by right-clicking within a cell with a missing value and choosing a value from **Value Labels > Select Value from Value Label 'repairs'**. This can be useful when a value label has many possible values.

6. We would now like to delete the trunk variable. We can do this by right-clicking on the trunk variable and selecting the **Drop Selected Data** menu item. Because this can lead to data loss,

the Data Editor asks whether we would like to drop the selected variable. Click on the **Yes** button.

7. To finish up, we would like to create a variable containing the gallons of gasoline per 100 miles driven for each of the cars.

    a. Click on the **Tools > Add Variable at End of Dataset...** menu item to bring up the *generate* dialog box.

    b. Type gp100m in the *Variable name* box.

    c. Being sure that the *Specify a value or an expression* radio button is selected, type 100/mpg in the formula box. We could have clicked the **Create...** button to open the *Expression builder* dialog box, but this formula was simple enough to type.

    d. Click on **OK**. You can scroll to the right to see the newly created variable.

Throughout this data editing, session we have been using the Data Editor to manipulate the data. If you look in the Results window, you will see the commands and their output. You can also see all the commands generated by the Data Editor in the Review window. If you wanted to save the editing commands to use again later, you could do the following:

1. Click on the last command that came from the Data Editor.
2. Scroll up until you find the edit command that originally opened the Data Editor.
3. While holding the *Shift* key down, click on the command following edit.
4. Right-click on one of the highlighted commands.
5. Select **Send to Do-file Editor**.

This will save all the commands you issued into the Do-file Editor. You could then save them as a do-file which you could run again later. We will talk more about the Do-file Editor in [GSU] **13 Using the Do-file Editor—automating Stata**. You can find help about do-files in [U] **16 Do-files**.

If you want to save this dataset, save it under a new name by using **File > Save As...** so as not to overwrite the original dataset.

Working with snapshots

The Data Editor allows you to save to disk *snapshots* of whatever dataset you are working on. These are temporary copies of the dataset—they will be deleted when you exit Stata, so they need to be treated as temporary. Still, there are many uses for snapshots, such as

- Saving a temporary copy of the data in memory so that another dataset can be opened and viewed.
- Saving stages of work, which can be recovered in case of doing something disastrous.
- Saving pieces of datasets while doing analyses.

We will keep using the auto dataset from above; if you are starting here, you can start fresh by typing sysuse auto in the Command window to open the dataset. (If you get a warning about data in memory being lost, either use clear or save your data. See [GSU] **5 Opening and saving Stata datasets** for more information.) If we open the Data Editor and click on **Tools > Snapshots...**, we see the following. If you are starting fresh, you will see numbers rather than labels for rep78.

(Continued on next page)

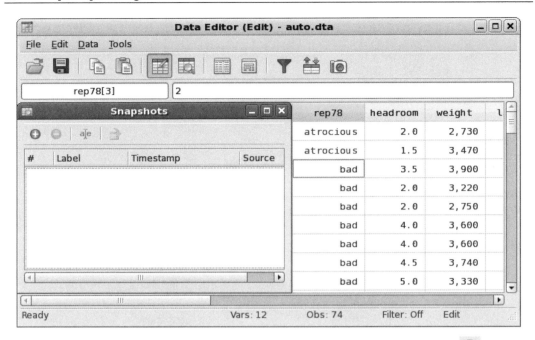

To begin with, only one button is active. Click the active button—the **Add** button, . It brings up a dialog asking for a label, or name, for the snapshot. Give it an inventive name, such as Start. You can see that a snapshot is now listed in the Snapshots window, and all the buttons are highlighted.

Add: Save a new snapshot with a timestamp and label.

Remove: Erase a snapshot. This deletes the temporary snapshot file, but does not affect the data in memory.

Change Label: Edit the label of the selected (highlighted) snapshot.

Restore: Replace the data in memory with the data from the selected snapshot. If the data in memory have changed, you will get a dialog box confirming your action.

You should now try manipulating the dataset using the tools we have seen. Once you have done this, create another snapshot, calling it Changed. Open the *Snapshots* window and restore the Start snapshot by clicking first on it and then on the **Restore** button to see where you started. You can then go back to where you were working by restoring your Changed snapshot.

Snapshots continue to be available until either they are deleted or you exit Stata. You can thus use snapshots of one dataset while working on another. You will find your own uses for snapshots—just take care to save datasets you want for future use.

Dates and the Data Editor

The Data Editor has two special tools for working with dates in Stata. To see these in action, we will need to open another dataset. Either save your dataset or clear it out, and then type sysuse sp500 in the Command window. Look in the Data Editor to see what you have.

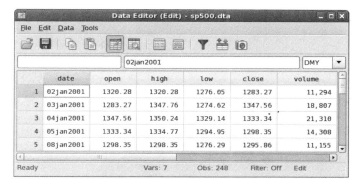

You can see that there is a `date` variable that has January 2, 2001, as its first day, but it is in Stata's default format for dates. You should also notice in the upper right corner, the *date mask* field, which shows `DMY`. This affects how dates are entered when editing data.

We will start with formatting:

1. Right-click within a cell in the `date` column, and select **Variable Properties...**.
2. In the *Variable Properties* window, click the button to the right of the *Format* field.
3. The *Create format* dialog tells us three pieces of information about the date format:
    a. These are daily dates. As you can see, Stata understands other types of dates that are often used in financial data.
    b. Looking at the bottom of the dialog, you can see that the Stata default date format is `%td`. This stands for "time-daily".
    c. This default format is displayed as, for example, `30apr2009`. (Of course, this was clear in the Data Editor.)
4. There are many pre-made date formats in the Samples pane. Click on `April 30, 2009`. You can see how the format would be specified at the bottom of the dialog.
5. Click on **OK** to close the *Create format* dialog.
6. Click on the **Apply** button to change the `date` format.

This is a very simple way to change date formats. For complete information on dates and date formats, see [D] **dates and times**.

We will now change some of the dates to illustrate how the date mask works. By default, the date mask is set to `DMY`. This means dates can be entered in many different fashions, as long as the order is day, month, year. Try the following:

1. With the `date[1]` cell as the active cell, type `3jan2009` and press the *Enter* key. Stata understands the order of the mask.
2. Enter `14012009` and press *Enter*. Stata still understands the mask, even though there are no separators.
3. Click within the *date mask* field, and choose `MDY` from the drop-down menu.
4. Click on any observation in the `date` column.
5. Type `March 13, 2001` and press *Enter*. Stata will still understand.

Working in this fashion is the fastest way to edit dates by hand. If you look in the Results window, you will see why.

We are now done with this dataset, so type `clear` and press *Enter*.

Data Editor advice

As you could see above, a small mistake in the Data Editor could cause large problems in your dataset. You really must take care in how you edit your data.

1. People who care about data integrity know that editors are dangerous—it is easy to accidentally make changes. **Never use the Data Editor in edit mode when you just want to look at your data.** Use the Data Editor in browse mode (or the `browse` command).

2. If you must edit your data, protect yourself by limiting the dataset's exposure. For example, if you need to change `rep78` only if it is missing, find a way to look at just the missing values for `rep78` and any other variables needed to make the change. This will make it impossible for you to change (damage) variables or observations other than those you view. We will explore this aspect shortly.

3. Even with these caveats, Stata's editor is safer than most because it records commands in the Results window. Use this feature to log your output and make a permanent record of the changes. Then, you can verify that the changes you made are the changes you wanted to make. See [GSU] **16 Saving and printing results by using logs** for information on creating log files.

Filtering and hiding

We would now like to investigate restricting our view of the data we see in the editor. This feature is useful for the reasons mentioned above, and as we will see, it helps if we would like to browse through the data of a large dataset. In any case, we would like to focus on *some*, not *all* of the data, be it some of the variables, some of the observations, or even just some observations within some variables. We will show you how this is done by using both the interface and commands.

Open the `auto` dataset by typing `sysuse auto`. Once you have done that, open the Data Editor.

Suppose that we would like to edit only those observations for which `rep78` is missing. We will need to look at the make of the car so that we know which observations we are working with, but we do not need to see any other variables. We will work as though we had a very large dataset to work with.

1. Click on the **Hide/Show Variables** button, .
2. Before we get started, try experimenting with filtering variables by entering some text into the *Enter filter text here* box. If you had a dataset with many variables, this would be a good way to reduce the number of variables visible when selecting your desired variables. When you are done, delete your filter text.
3. Right-click on any variable and select **Select All** from the very short contextual menu.
4. Click on any checkbox to deselect all the variables.
5. Click on the `make` variable to select it and deselect all the other variables.
6. Click on the checkbox for `make`.
7. Click on the checkbox for `rep78`.
8. Close the *Hide/Show Variables* window.

When you bring the Data Editor to the front, you will see that the only variables visible are `make` and `rep78`. If you look in the Command window, no commands have been issued, because hiding the variables does not affect the dataset—it affects only what shows in the Data Editor.

We now have protected ourselves by using only those variables that we need. We should now reduce our view to only those observations for which `rep78` is missing. This is simple.

1. Click on the **Filter Observations** button, .
2. Enter `missing(rep78)` in the *Filter by expression* field.

3. Click on the **Apply Filter** button.

4. If you are curious, click the **...** button. It opens up an *Expression builder* dialog box. This lists the wide variety of functions available in Stata. See [D] **functions**.

Now we are focused on the part of the dataset in which we would like to work, and we cannot destroy or mistakenly alter other data by stray keystrokes in the Data Editor window.

It is worth learning how to hide variables and filter observations in the Data Editor from the Command window. This can be quite convenient if you are going to restrict your view, as we did above. To work from the Command window, we must use the `edit` command together with a varlist (variable list) along with `if` and `in` qualifiers in the Command window. By using a varlist, we restrict the variables we look at, whereas the `if` and `in` qualifiers restrict the observations we see. ([GSU] **10 Listing data and basic command syntax** contains many examples of using a command with a variable list and `if` and `in`.) Suppose we want to correct the missing values for `rep78`. The minimum amount of data we need to expose are `make` and `rep78`. To see this minimal amount of information and hence to minimize our exposure to making mistakes, we enter the commands

```
. sysuse auto
(1978 Automobile Data)
. edit make rep78 if missing(rep78)
```

and we would see following screen:

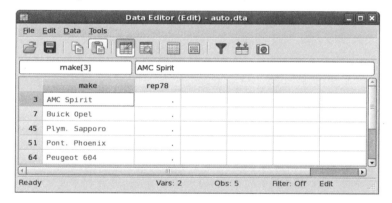

Once again, we are safe and sound.

Keep this lesson in mind if you edit your data. It is a lesson well learned.

Browse mode

The purpose of using the Data Editor in browse mode is to be able to look at data without altering it by stray keystrokes. You can start the Data Editor in browse mode by clicking on the **Data Editor (Browse)** button, [image], or by typing `browse` in the Command window. When working in browse mode, all contextual menu items that would let you alter the data, the labels, or any of the display formats for the variables are disabled. You may view a variable's properties via the **Variable Properties...** menu item, but you may not make any changes. You still can filter observations and hide variables to get a restricted view, because these actions do not change the dataset.

Note: Because you can still use Stata menus not related to the Data Editor and because you can still type commands in the Commands window, it is possible to change the data even if the Data Editor is in browse mode. In fact, this means you can watch how your commands affect the dataset. You are merely restricted from using the Data Editor itself to change the data.

.

7 Using the Variables Manager

The Variables Manager in Stata(GUI)

This chapter discusses the Variables Manager for Stata (GUI). Stata(console) users need to consult the *Data-Management Reference Manual* for details concerning the commands issued by the Variables Manager.

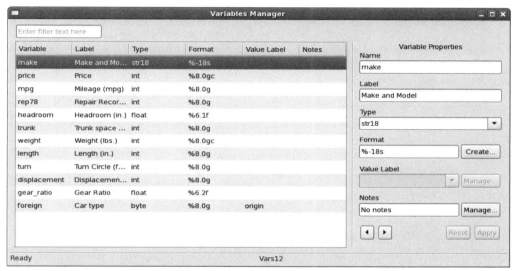

The Variables Manager is a tool for managing properties of variables both individually and in groups. It can be used to create variable and value labels, rename variables, change display formats, and manage notes. It has the ability to filter and group variables as well as to create variable lists. Users will find these features useful for managing large datasets.

Any action you take in the Variables Manager results in a command being issued to Stata as though you had typed it in the Command window. This means that you can keep good records and learn commands by using the Variables Manager.

The Variable pane

The Variable pane shows the list of variables in the dataset. This list can be manipulated in a variety of ways.

- The variables can be filtered by entering text into the filter box in the upper left corner. This can be a good way to zoom in on similarly named or labeled variables.
- The list can be sorted by clicking on the columns.
  a. If you click on a column, it will sort in ascending order.
  b. A second click on the same column will change to sorting in descending order.
  c. To restore the sort order to the original sort order, right-click within the variable list, and select **Show Variables in Dataset Order**.

The sort order affects only the Variable Manager window—the dataset itself stays the same.

Right-clicking on the Variable pane

Right-clicking on the Variable pane displays a menu from which you can do many common tasks:

- **Keep Selected Variables** to keep only the selected variables in the dataset, dropping all the others.
- **Drop Selected Variables** to drop all the selected variables from the dataset.
- **Manage Notes for Selected Variable...** to open a window that allows adding and deleting notes for a single variable. This is disabled if multiple variables are selected.
- **Manage Notes for Dataset...** to open a window that allows adding and deleting notes for the dataset as a whole.
- **Copy Varlist** to copy the names of the selected variables to the Clipboard.
- **Select All** to select all visible variables. If a variable has become hidden because of the filter, it will not be selected.
- **Send Varlist to Command Window** to insert the names of the selected variables in the Command window. Combined with grouping and sorting, this can be a useful way to create variable lists in large datasets.

The Variable Properties pane

The Variable Properties pane can be used to manipulate the properties of variables selected in the Variable pane. With one variable selected, you can manipulate all properties of the variable. With many variables selected, you can change their formats or types, as well as assign value labels all at once. These fields work in the same fashion as those shown in **Renaming and formatting variables** in [GSU] **6 Using the Data Editor**. We can also manage the notes Stata allows you to attach to variables and the dataset—we will show an example below.

Managing notes

Stata allows you to attach notes to both variables and the dataset as a whole. These are simple text notes that you can use to document whatever you like—the source of the dataset, data collection quirks associated with a variable, what you need to investigate about a variable, or anything else.

Start by selecting a variable in the Variable pane. We will work with the `price` variable. Click on the **Manage...** button next to the *Notes* field, and you will see the following dialog appear:

We will add a few notes:

1. Click on the **Add** button to add a note.

2. Type TS - started working. TS with a trailing space inserts a timestamp in the note.

3. Add two more notes. We added two notes about prices:

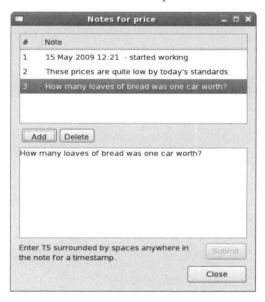

It is worth experimenting with adding, deleting, and editing notes. These can be an invaluable memory aid when working on projects that last over a long period of time. Any time you manipulate notes in the Notes Manager, you create Stata commands.

Notes

8 Importing data

Copying and pasting in Stata(GUI)

One of the easiest ways to get data into Stata is often overlooked: you can copy data from most applications that understand the concept of a *table* and then paste the data into the Data Editor. This approach works for all spreadsheet applications, many database applications, some word-processing applications, but few web browsers. Just copy the full range of data, paste it into the Data Editor, and everything will probably work well. You can even copy a text file that has the pieces of data separated by commas and then paste it into the Data Editor. There is one warning, however: be sure that your spreadsheet does not contain blank rows, blank columns, repeated headers, or merged cells because these can cause trouble. As long as your spreadsheet looks like a table, you will be fine.

If you cannot simply copy and paste a dataset, this chapter is for you. If you would like to learn methods that lend themselves better to repetitive tasks, this chapter is also for you. If your friend has interesting data in a spreadsheet application that you do not own, this chapter is for you.

Commands for importing data

This chapter explains the various commands that Stata has for importing data. The three main commands for reading plain text (ASCII) are

- `insheet`, which is made for reading text files created by spreadsheet or database programs;
- `infile`, which is made for reading simple data that are separated by spaces, or rigidly formatted data aligned in columns; and
- `infix`, which is made for data aligned in columns but possibly split across rows.

Stata has other commands that can read other types of files and can even get data from external databases without the need for an interim file:

- The `fdause` command can read any SAS XPORT file, so data can be transferred from SAS to Stata in this fashion.
- The `odbc` command can be used to pull data directly from any data sources for which you have ODBC drivers.
- The `xmluse` command can read some XML files (most notably, Microsoft Excel's SpreadsheetML).

Each command expects the file that they are reading to be in a specific format. This chapter will explain some of those formats and give some examples. For the full story, consult the *Data-Management Reference Manual*.

(Continued on next page)

The insheet command

The insheet command was specially developed to read in text (ASCII) files that were created by spreadsheet programs. Many people have data that they enter into their spreadsheet programs. All spreadsheet programs have an option to save the dataset as a text (ASCII) file with the columns delimited with either tab characters or commas. Some spreadsheet programs also save the column titles (variable names in Stata) in the text file.

To read in this file, you have only to type insheet using *filename*, where *filename* is the name of the text file. The insheet command will determine whether there are variable names in the file, what the delimiter character is (tab or comma), and what type of data is in each column. If *filename* contains spaces, put double quotes around the filename.

If you have a text (ASCII) file that you created by saving data from a spreadsheet program, then try the insheet command to read that data into Stata.

By default, the insheet command understands files that use the tab or comma as the column delimiter. If you have a file that uses spaces as the delimiter, use the infile command instead. See the infile section of this chapter for more information. If you have a file that uses another character as the delimiter, use insheet's delimiter() option; see [D] **insheet** for more information.

Suppose that your friend sent you the file sample.csv. It was saved from a spreadsheet application by saving the data as a *comma-separated–values* (CSV) file. It contains the following lines:

```
VW Rabbit,4697,25,1930,3.78
Olds 98,8814,21,4060,2.41
Chev. Monza,3667,,2750,2.73
,4099,22,2930,3.58
Datsun 510,5079,24,2280,3.54
Buick Regal,5189,20,3280,2.93
Datsun 810,8129,,2750,3.55
```

These data correspond to the make, price, MPG, weight, and gear ratio of a few very old cars. The variable names are not in the file (so insheet will assign its own names), and the fields are separated by a comma. Use insheet to read the data:

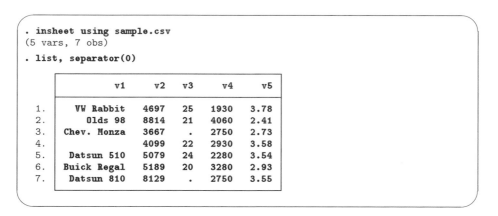

```
. insheet using sample.csv
(5 vars, 7 obs)

. list, separator(0)

            v1      v2    v3     v4      v5

  1.   VW Rabbit   4697    25   1930    3.78
  2.     Olds 98   8814    21   4060    2.41
  3.  Chev. Monza  3667     .   2750    2.73
  4.               4099    22   2930    3.58
  5.   Datsun 510  5079    24   2280    3.54
  6.  Buick Regal  5189    20   3280    2.93
  7.   Datsun 810  8129     .   2750    3.55
```

If you want to specify better variable names, you can include the desired names in the command:

```
. insheet make price mpg weight gear_ratio using sample.csv
(5 vars, 7 obs)
. list, separator(0)

           make    price   mpg   weight   gear_r~o

  1.    VW Rabbit    4697    25     1930     3.78
  2.      Olds 98    8814    21     4060     2.41
  3.   Chev. Monza   3667     .     2750     2.73
  4.                 4099    22     2930     3.58
  5.    Datsun 510   5079    24     2280     3.54
  6.   Buick Regal   5189    20     3280     2.93
  7.   Datsun 810    8129     .     2750     3.55
```

As a side note about displaying data, Stata listed gear_ratio as gear_r~o in the output from list. gear_r~o is a unique abbreviation for the variable gear_ratio. Stata displays the abbreviated variable name when variable names are longer than eight characters.

To prevent Stata from abbreviating gear_ratio, you could specify the abbreviate(10) option:

```
. list, separator(0) abbreviate(10)

           make    price   mpg   weight   gear_ratio

  1.    VW Rabbit    4697    25     1930     3.78
  2.      Olds 98    8814    21     4060     2.41
  3.   Chev. Monza   3667     .     2750     2.73
  4.                 4099    22     2930     3.58
  5.    Datsun 510   5079    24     2280     3.54
  6.   Buick Regal   5189    20     3280     2.93
  7.   Datsun 810    8129     .     2750     3.55
```

For more information on the ~ abbreviation and on list, see [GSU] **10 Listing data and basic command syntax**.

For this simple example, you could have copied the above file and pasted it into the Data Editor, because the Data Editor also understands the comma as the delimiter.

(Continued on next page)

infile with unformatted data

The file afewcars.raw contains the following lines, which we can view using the type command:

```
. type afewcars.raw
"VW Rabbit"      4697     25      1930     3.78
"Olds 98"       8814     21      4060    2.41
"Chev. Monza"   3667     .       2750    2.73
""              4099     22      2930    3.58
"Datsun 510"    5079     24      2280    3.54
"Buick Regal"
5189
20
3280
2.93
"Datsun 810"    8129     ***     2750    3.55
```

These correspond to the make, price, MPG, weight, and gear ratio of a few cars.

This file is a mess. It uses spaces, instead of a good special character, to delimit pieces of data. The second line is not formatted the same as the first line. The sixth car's data are spread over five lines. There are some data missing, and the missing values have different representations: the MPG on the third line has a '.', the make on the fourth line contains "", and the MPG on the last line contains '***'. The one redeeming feature of the file is that the makes of cars are quoted, so that spaces in the names, such as "Olds 98", will not make the file unreadable. The infile command can deal with such a messy file.

```
. infile str18 make price mpg weight gear_ratio using afewcars
'***' cannot be read as a number for mpg[7]
(7 observations read)
```

Here are some notes about this example:

- The phrase using afewcars in the command is interpreted by Stata to mean using afewcars.raw. The extension .raw is assumed by infile if you do not specify otherwise. If the data had been stored in afewcars.txt, you would have had to specify using afewcars.txt.

- infile does not require each observation to be on one line. The location of line breaks does not matter, because they are treated the same as a space (the computer term for this is that they are treated as *whitespace*).

- The variable names were required in this example so that infile could know how many variables there were. It could not have used the number of pieces of data on the first line, because it is not treating line breaks as special.

- Entries that are not understood (such as ***) are mentioned and are stored as missing values.

- Entries that use Stata's conventions for missing values are not mentioned. There are two examples of this: '.', indicating a numeric missing value, and "", indicating a string missing value. However, with infile you must explicitly have a "" in your file so that it does not read the next number in the file as the value of this string variable.

infile with fixed-format data

File `cars2.raw` contains the following lines:

```
. type cars2.raw
VW Rabbit          4697        25        1930      3.78
Olds 98            8814        21        4060      2.41
Chev. Monza        3667                  2750      2.73
                   4099        22        2930      3.58
Datsun 510         5079        24        2280      3.54
Buick Regal        5189        20        3280      2.93
Datsun 810         8129                  2750      3.55
```

These data are simple for a human to read, because they are formatted nicely, but much more difficult for the computer to read without guidance. There are two related reasons for this: there are no double quotes around the strings in the first column and they include blanks, and there are blanks in the first and third columns when the data are missing. This means that `infile` cannot use whitespace as before to know where one piece of data ends and the next begins. Trying to separate the data by using whitespace leads to a complete failure, because it assigns each chunk of text to each variable in turn:

make is	VW	in the first observation	
price is	Rabbit	in the first observation	(error, stores as missing value)
MPG is	4697	in the first observation	(wrong)
weight is	25	in the first observation	(also wrong)
gear ratio is	1930	in the first observation	(wrong again)
make is	3.78	in the second observation	(stores "3.78" as a string)
price is	Olds	in the second observation	(error, stores as missing value)
MPG is	98	in the second observation	(surprise)
⋮	⋮	⋮	

This problem is referred to as *loss of synchronization*.

We could solve this problem if we told `infile` that the data were nicely aligned in columns. This alignment can be passed on to `infile` in the form of a *dictionary file*. Here is a dictionary file that would work for these data. We can create it by using the Do-file Editor and save it in the file `forcars2.dct`:

```
. type forcars2.dct
dictionary using cars2.raw {
        _column(1)        str18 make        %11s
        _column(19)       price             %4f
        _column(31)       mpg               %2f
        _column(39)       weight            %4f
        _column(47)       gear_ratio        %4f
}
```

You can see that the column numbers tell `infile` where to start reading each variable. Furthermore, the dictionary contains the name of the file for which it was made. The details of constructing dictionaries are described in [D] **infile (fixed format)**. To read the data, type

```
. infile using forcars2
dictionary using cars2.raw {
        _column(1)       str18 make        %11s
        _column(19)      price             %4f
        _column(31)      mpg               %2f
        _column(39)      weight            %4f
        _column(47)      gear_ratio        %4f
}
(7 observations read)
. list
```

	make	price	mpg	weight	gear_r~o
1.	VW Rabbit	4697	25	1930	3.78
2.	Olds 98	8814	21	4060	2.41
3.	Chev. Monza	3667	.	2750	2.73
4.		4099	22	2930	3.58
5.	Datsun 510	5079	24	2280	3.54
6.	Buick Regal	5189	20	3280	2.93
7.	Datsun 810	8129	.	2750	3.55

The infile command uses a dictionary because no variables are specified. It adds a file extension, .dct, when no extension is specified.

Stata has a second command, infix, for reading fixed-format files. Its full documentation is in [D] **infix (fixed format)**.

Importing files from other software

Stata has some more specialized methods for reading data that were created or stored in another format. These are useful when you do not have control over the format in which the data are distributed.

The fdause command can read and create SAS XPORT Transport files. See fdause and fdasave in [D] **fdasave** for full details.

If you have software that supports ODBC (*Open Database Connectivity*), you can read data by using the odbc command without the need to create interim files. See [D] **odbc** for full details. The FAQ for setting up ODBC is also helpful; view it at http://www.stata.com/support/faqs/data/odbcmu.html.

If you would like to move data by using XML (*Extensible Markup Language*), Stata has the xmluse and xmlsave commands available. See xmluse and xmlsave in [D] **xmlsave** for full details.

Here is a brief summary of the choices:
- If you have a table, you could try copying it and pasting into the Data Editor.
- If you have a file exported from a spreadsheet or database application to a tab-delimited or CSV file, use insheet.
- If you have a free-format, space-delimited file, use infile with variable names.
- If you have a fixed-format file, either use infile with a dictionary or use infix.
- If you have a SAS XPORT file, use fdause.
- If you have a database accessible via ODBC, use odbc.
- If you have an XML file, use xmluse.
- Finally, you can purchase a transfer program that will convert the other software's data file format to Stata's data file format. See [U] **21.4 Transfer programs**.

9 Labeling data

Making data readable

This chapter discusses, in brief, labeling of the dataset, variables, and values. Such labeling is critical to careful use of data. Labeling variables with descriptive names clarifies their meanings. Labeling values within categorical variables keeps the true meanings of the categories from becoming obscure. These points are crucial when sharing data with others, including your future self. Labels are also used in the output of most Stata commands, so proper labeling of the dataset will make much more readable results. We will work through an example of properly labeling a dataset, its variables, and the values of one encoded variable.

The dataset structure: The describe command

In [GSU] **8 Importing data**, we saved a dataset called afewcars.dta. We will get this dataset into a shape that a colleague would understand. Let's see what it contains.

```
. use afewcars
. list, separator(0)

          make     price    mpg    weight    gear_r~o

 1.    VW Rabbit     4697     25      1930        3.78
 2.      Olds 98     8814     21      4060        2.41
 3.   Chev. Monza    3667      .      2750        2.73
 4.                  4099     22      2930        3.58
 5.    Datsun 510    5079     24      2280        3.54
 6.   Buick Regal    5189     20      3280        2.93
 7.    Datsun 810    8129      .      2750        3.55
```

The data allow us to make some guesses at the values in the dataset, but, for example, we do not know the units in which the price or weight is measured, and the term "mpg" could be confusing for people outside the United States. Perhaps we can learn something from the description of the dataset. Stata has the aptly named describe command for this purpose (as we saw in [GSU] **1 Introducing Stata—sample session**).

(Continued on next page)

71

```
. describe
Contains data from afewcars.dta
  obs:            7
  vars:           5                                11 May 2009 20:47
  size:         294 (99.9% of memory free)

                storage  display    value
variable name   type     format     label      variable label

make            str18    %18s
price           float    %9.0g
mpg             float    %9.0g
weight          float    %9.0g
gear_ratio      float    %9.0g

Sorted by:
```

Though there is precious little information that could help us as a researcher, we can glean some information here about how Stata thinks of the data from the first three columns of the output.

1. The *variable name* is the name we use to tell Stata about a variable.

2. The *storage type* (otherwise known as the *data type*) is the way in which Stata stores the data in a variable. There are six different storage types:

   a. For integers:

   > byte for integers between -127 and 100
   >
   > int for integers between $-32{,}767$ and $32{,}740$
   >
   > long for integers between $-2{,}147{,}483{,}647$ and $2{,}147{,}483{,}620$

   b. For real numbers:

   > float for real numbers with 8.5 digits of precision
   >
   > double for real numbers with 16.5 digits of precision

   c. For strings (text) from 1–244 characters:

   > str1 for one-character-long strings
   >
   > str2 for two-character-long strings
   >
   > str3 for three-character-long strings
   >
   > . . .
   >
   > str244 for 244-character-long strings

Storage types affect both the precision of computations and the size of datasets. A quick guide to storage types is available at help datatype or in [D] **data types**.

3. The *display format* controls how the variable is displayed; see [U] **12.5 Formats: Controlling how data are displayed**. By default, Stata sets it to something reasonable given the storage type. We would like to make this dataset into something containing all the information we need.

To see what a well-labeled dataset looks like, we can take a look at a dataset stored at the Stata Press repository. We need not load the data (and disturb what we are doing); we do not even need a copy of the dataset on our machine. (You will learn more about Stata's Internet capabilities in [GSU] **19 Updating and extending Stata—Internet functionality**.) All we need to do is direct describe to look at the proper file by using the command describe using *filename*.

```
. describe using http://www.stata-press.com/data/r11/auto
Contains data                                  1978 Automobile Data
  obs:            74                            13 Apr 2009 17:45
  vars:           12
  size:        3,478

              storage   display    value
variable name   type    format     label       variable label

make            str18   %-18s                   Make and Model
price           int     %8.0gc                  Price
mpg             int     %8.0g                   Mileage (mpg)
rep78           int     %8.0g                   Repair Record 1978
headroom        float   %6.1f                   Headroom (in.)
trunk           int     %8.0g                   Trunk space (cu. ft.)
weight          int     %8.0gc                  Weight (lbs.)
length          int     %8.0g                   Length (in.)
turn            int     %8.0g                   Turn Circle (ft.)
displacement    int     %8.0g                   Displacement (cu. in.)
gear_ratio      float   %6.2f                   Gear Ratio
foreign         byte    %8.0g      origin       Car type

Sorted by:  foreign
```

This output is much more informative. There are three locations where labels are attached that help explain what the dataset contains:

1. In the first line, 1978 Automobile Data is the *data label*. It gives information about the contents of the dataset. Data can be labeled by selecting **Data > Data utilities > Label utilities > Label dataset** or by using the label data command.

2. There is a *variable label* attached to each variable. Variable labels are not only how we would refer to the variable in normal, everyday conversation, but they also contain information about the units of the variables. Variables can be labeled by right-clicking on the variable in the Variables window and selecting **Edit Variable Label for '*varname*'...**, by using the Variables Manager, or by using the label variable command.

3. The foreign variable has an attached *value label*. Value labels allow numeric variables, such as foreign, to have words associated with numeric codes. The describe output tells you that the numeric variable foreign has value label origin associated with it. Although not revealed by describe, the variable foreign takes on the values 0 and 1, and the value label origin associates 0 with Domestic and 1 with Foreign. If you browse the data (see [GSU] **6 Using the Data Editor**), foreign appears to contain the values "Domestic" and "Foreign". The values in a variable are labeled in two stages. The value label must first be defined. This can be done in the Data Editor, the Variables Manager, by selecting **Data > Data utilities > Label utilities > Manage value labels** or by typing the label define command. After the labels have been defined, they must be attached to the proper variables either by selecting **Data > Data utilities > Label utilities > Assign value label to variables** or by using the label values command. One note: It is not necessary for the value label to have a name different from that of the variable. You could just as easily have used a value label named foreign.

(Continued on next page)

Labeling datasets and variables

We will now take our `afewcars.dta` dataset and give it proper labels. We will do this via the Command window, because it is simple to do in this fashion. If you use the menus and dialogs, the Data Editor, the Variables Manager, or any other GUI element, you will end up with the same commands in your log.

```
. use afewcars
. describe
Contains data from afewcars.dta
  obs:           7
  vars:          5                              11 May 2009 20:47
  size:        294 (99.9% of memory free)

              storage   display    value
variable name   type    format     label       variable label

make          str18     %18s
price         float     %9.0g
mpg           float     %9.0g
weight        float     %9.0g
gear_ratio    float     %9.0g

Sorted by:
. label data "A few 1978 cars"
. label variable make "Make and Model"
. label variable price "Price (USD)"
. label variable mpg "Mileage (mile per gallon)"
. label variable weight "Vehicle weight (lbs.)"
. label variable gear_ratio "Gear Ratio"
. describe
Contains data from afewcars.dta
  obs:           7                              A few 1978 cars
  vars:          5                              11 May 2009 20:47
  size:        294 (99.9% of memory free)

              storage   display    value
variable name   type    format     label       variable label

make          str18     %18s                    Make and Model
price         float     %9.0g                   Price (USD)
mpg           float     %9.0g                   Mileage (mile per gallon)
weight        float     %9.0g                   Vehicle weight (lbs.)
gear_ratio    float     %9.0g                   Gear Ratio

Sorted by:
. save, replace
file afewcars.dta saved
```

Warning: When you change or define labels on a dataset in memory, it is worth saving the dataset right away. Because the actual data in the dataset did not change, Stata will not prevent you from exiting or loading a new dataset later, and you will lose your labels.

Labeling values of variables

Suppose that we now added a new indicator variable to the dataset that is 0 if the car was made in the United States and 1 if it was foreign made. Our data would now look like this:

```
. list, separator(0)

              make    price    mpg   weight   gear_r~o   foreign

   1.    VW Rabbit     4697     25     1930       3.78         1
   2.       Olds 98    8814     21     4060       2.41         0
   3.   Chev. Monza    3667      .     2750       2.73         0
   4.                  4099     22     2930       3.58         0
   5.    Datsun 510    5079     24     2280       3.54         1
   6.   Buick Regal    5189     20     3280       2.93         0
   7.    Datsun 810    8129      .     2750       3.55         1
```

Though the definitions of the categories "0" and "1" are clear in this context, it still would be worthwhile to give the values explicit labels, because it will make output clear to people who are not so familiar with antique automobiles. This is done with a value label.

We saw an example of creating and attaching a value label by using the point-and-click interface available in the Data Editor in **Changing data** in [GSU] **6 Using the Data Editor**. Here we'll do it directly from the Command window.

```
. label define origin 0 "domestic" 1 "foreign"
. label values foreign origin
. describe
Contains data from afewcars.dta
  obs:             7                          A few 1978 cars
 vars:             6                          11 May 2009 20:47
 size:           301 (99.9% of memory free)

               storage   display    value
variable name    type    format     label      variable label

make            str18    %18s                   Make and Model
price           float    %9.0g                  Price (USD)
mpg             float    %9.0g                  Mileage (mile per gallon)
weight          float    %9.0g                  Vehicle weight (lbs.)
gear_ratio      float    %9.0g                  Gear Ratio
foreign         byte     %8.0g      origin

Sorted by:
     Note:  dataset has changed since last saved
. save, replace
file afewcars.dta saved
```

From this, we can see that a value label is defined via

label define *labelname* # "*contents*" # "*contents*" ...

It can then be attached to a variable via

label values *variablename labelname*

Once again, we need to save the dataset to be sure that we do not mistakenly lose the labels later.

If you had wanted to define the value labels by using a point-and-click interface, the easiest way to do that would be to open the Variables Manager by clicking on the **Variables Manager** button, ▦ , and then manage labels. See [GSU] **7 Using the Variables Manager** for more information.

There is more to value labels than what was covered here; see [U] **12.6.3 Value labels** for a complete treatment.

You may also add notes to your data and your variables. This feature was discussed in **Managing notes** in [GSU] **7 Using the Variables Manager**. You can learn more about notes by typing `help notes` or you can get the full story in [D] **notes**.

10 Listing data and basic command syntax

Command syntax

This chapter gives a basic lesson on Stata's command syntax while showing how to control the appearance of a data list.

As we have seen throughout this manual, you have a choice between using menus and dialogs and using the Command window. Although many find the menus more natural and the Command window baffling at first, some practice makes working via the Command window often much faster than using menus and dialogs. The Command window can become a faster way of working because of the clean and regular syntax of Stata commands. We will cover enough to get you started; help language has more information and examples, and [U] **11 Language syntax** has all the details.

The syntax for the list command can be seen by typing help list:

$$\underline{\text{list}} \; [\textit{varlist}] \; [\textit{if}] \; [\textit{in}] \; [, \textit{options}]$$

Here is how to read this syntax:
- Anything inside square brackets is optional. For the list command,
    a. *varlist* is optional. A *varlist* is a list of variable names.
    b. *if* is optional. The if qualifier restricts the command to run only on those observations for which the qualifier is true. We saw examples of this in [GSU] **6 Using the Data Editor**.
    c. *in* is optional. The in qualifier restricts the command to run on particular observation numbers.
    d. , and *options* are optional. *options* are separated from the rest of the command by a comma.
- Optional pieces do not preclude one another unless explicitly stated. For the list command, it is possible to use a *varlist* with *if* and *in*.
- If a part of a word is underlined, the underlined part is the minimum abbreviation. Any abbreviation at least this long is acceptable.
    a. The l in list is underlined, so l, li, and lis are all equivalent to list.
- Anything not inside square brackets is required. For the list command, only the command itself is required.

Keeping these rules in mind, let's investigate how list behaves when called with different arguments. We will be using the dataset afewcars.dta from the end of the previous chapter.

list with a variable list

Variable lists (or *varlist*s) can be specified in a variety of ways, all designed to save typing and encourage good variable names.
- The *varlist* is optional for list. This means that if no variables are specified, it is equivalent to specifying all variables. Another way to think of it is that the default behavior of the command is to run on all variables, unless restricted by a *varlist*.
- You can list a subset of variables explicitly, as in list make mpg price.
- There are also many shorthand notations:
    m* means all variables starting with m.
    price-weight means all variables from price through weight in dataset order.
    ma?e means all variables starting with ma, followed by any character, and ending in e.

- You can list a variable by using an abbreviation unique to that variable, as in `list gear_r~o`. If the abbreviation is not unique, Stata returns an error message.

```
. list

                    make    price    mpg   weight   gear_r~o    foreign

  1.           VW Rabbit     4697     25     1930       3.78      foreign
  2.             Olds 98     8814     21     4060       2.41     domestic
  3.         Chev. Monza     3667      .     2750       2.73     domestic
  4.                        4099      22     2930       3.58     domestic
  5.          Datsun 510     5079     24     2280       3.54      foreign

  6.         Buick Regal     5189     20     3280       2.93     domestic
  7.          Datsun 810     8129      .     2750       3.55      foreign

. l make mpg price

                    make    mpg    price

  1.           VW Rabbit     25     4697
  2.             Olds 98     21     8814
  3.         Chev. Monza      .     3667
  4.                         22     4099
  5.          Datsun 510     24     5079

  6.         Buick Regal     20     5189
  7.          Datsun 810      .     8129

. list m*

                    make    mpg

  1.           VW Rabbit     25
  2.             Olds 98     21
  3.         Chev. Monza      .
  4.                         22
  5.          Datsun 510     24

  6.         Buick Regal     20
  7.          Datsun 810      .

. li price-weight

           price    mpg   weight

  1.        4697     25     1930
  2.        8814     21     4060
  3.        3667      .     2750
  4.        4099     22     2930
  5.        5079     24     2280

  6.        5189     20     3280
  7.        8129      .     2750
```

```
. list ma?e

            make

 1.    VW Rabbit
 2.      Olds 98
 3.   Chev. Monza
 4.
 5.   Datsun 510

 6.   Buick Regal
 7.   Datsun 810

. l gear_r~o

       gear_r~o

 1.       3.78
 2.       2.41
 3.       2.73
 4.       3.58
 5.       3.54

 6.       2.93
 7.       3.55
```

list with if

The `if` qualifier uses a logical expression to determine which observations to use. If the expression is true, the observation is used in the command; otherwise, it is skipped. The operators whose results are either true or false are

<	less than
<=	less than or equal
==	equal
>	greater than
>=	greater than or equal
!=	not equal
&	and
|	or
!	not (logical negation; ~ can also be used)
()	parentheses are for grouping to specify order of evaluation

In the logical expressions, & is evaluated before | (similar to multiplication before addition in arithmetic). You can use this in your expressions, but it is often better to use parentheses to ensure that the expressions are evaluated properly. See [U] **13.2 Operators** for the complete story.

```
. list

              make    price    mpg   weight   gear_r~o    foreign

   1.      VW Rabbit    4697     25     1930       3.78     foreign
   2.        Olds 98    8814     21     4060       2.41    domestic
   3.    Chev. Monza    3667      .     2750       2.73    domestic
   4.                   4099     22     2930       3.58    domestic
   5.     Datsun 510    5079     24     2280       3.54     foreign

   6.    Buick Regal    5189     20     3280       2.93    domestic
   7.     Datsun 810    8129      .     2750       3.55     foreign

. list if mpg > 22

              make    price    mpg   weight   gear_r~o    foreign

   1.      VW Rabbit    4697     25     1930       3.78     foreign
   3.    Chev. Monza    3667      .     2750       2.73    domestic
   5.     Datsun 510    5079     24     2280       3.54     foreign
   7.     Datsun 810    8129      .     2750       3.55     foreign

. list if (mpg > 22) & !missing(mpg)

              make    price    mpg   weight   gear_r~o    foreign

   1.      VW Rabbit    4697     25     1930       3.78     foreign
   5.     Datsun 510    5079     24     2280       3.54     foreign

. list make mpg price gear if (mpg > 22) | (price > 8000 & gear < 3.5)

              make     mpg   price   gear_r~o

   1.      VW Rabbit     25    4697       3.78
   2.        Olds 98     21    8814       2.41
   3.    Chev. Monza      .    3667       2.73
   5.     Datsun 510     24    5079       3.54
   7.     Datsun 810      .    8129       3.55

. list make mpg if mpg <= 22 in 2/4

           make    mpg

   2.    Olds 98     21
   4.                22
```

In the listings above, we see more examples of Stata treating missing numerical values as large values, as well as the care that should be taken when the if qualifier is applied to a variable with missing values. See [GSU] **6 Using the Data Editor**.

list with if, common mistakes

Here is a series of listings with common errors and their corrections. See if you can find the errors before reading the correct entry.

```
. list

               make    price    mpg   weight   gear_r~o    foreign
    1.      VW Rabbit    4697     25     1930       3.78    foreign
    2.        Olds 98    8814     21     4060       2.41   domestic
    3.    Chev. Monza    3667      .     2750       2.73   domestic
    4.                   4099     22     2930       3.58   domestic
    5.     Datsun 510    5079     24     2280       3.54    foreign

    6.    Buick Regal    5189     20     3280       2.93   domestic
    7.     Datsun 810    8129      .     2750       3.55    foreign

. list if mpg=21
=exp not allowed
r(101);
```

The error arises because "equal" is expressed by ==, not by =. Corrected, it becomes

```
. list if mpg==21

          make    price    mpg   weight   gear_r~o    foreign
    2.   Olds 98    8814     21     4060       2.41   domestic
```

Other common errors with logic:

```
. list if mpg==21 if weight > 4000
invalid syntax
r(198);
. list if mpg==21 and weight > 4000
invalid 'and'
r(198);
```

Joint tests are specified with &, not with the word and or multiple ifs. The if qualifier should be if mpg==21 & weight>4000, not if mpg==21 if weight>4000. Here is its correction:

```
. list if mpg==21 & weight > 4000

          make    price    mpg   weight   gear_r~o    foreign
    2.   Olds 98    8814     21     4060       2.41   domestic
```

A problem with string variables:

```
. list if make==Datsun 510
Datsun not found
r(111);
```

Strings must be in double quotes, as in make=="Datsun 510". Without the quotes, Stata thinks that Datsun is a variable that it cannot find. Here is the correction:

```
. list if make=="Datsun 510"

            make    price    mpg    weight    gear_r~o    foreign

  5.    Datsun 510    5079     24     2280        3.54    foreign
```

Confusing value labels with strings:

```
. list if foreign=="domestic"
type mismatch
r(109);
```

Value labels look like strings but the underlying variable is numeric. Variable foreign takes on values 0 and 1 but has the value label that attaches 0 to "domestic" and 1 to "foreign" (see [GSU] **9 Labeling data**). To see the underlying numeric values of variables with labeled values, use the label list command (see [D] **label**), or investigate the variable with codebook *varname*. We can correct the error here by looking for observations where foreign==0.

There is a second construction that also allows the use of the value label directly.

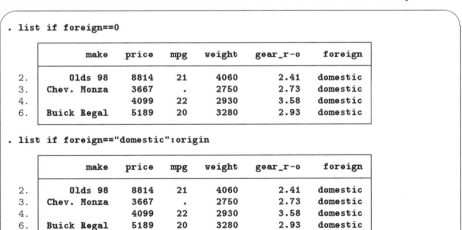

```
. list if foreign==0

            make    price    mpg    weight    gear_r~o    foreign

  2.        Olds 98    8814     21     4060        2.41    domestic
  3.    Chev. Monza    3667      .     2750        2.73    domestic
  4.                   4099     22     2930        3.58    domestic
  6.    Buick Regal    5189     20     3280        2.93    domestic

. list if foreign=="domestic":origin

            make    price    mpg    weight    gear_r~o    foreign

  2.        Olds 98    8814     21     4060        2.41    domestic
  3.    Chev. Monza    3667      .     2750        2.73    domestic
  4.                   4099     22     2930        3.58    domestic
  6.    Buick Regal    5189     20     3280        2.93    domestic
```

list with in

The in qualifier uses a *numlist* to give a range of observations that should be listed. *numlists* have the form of one number or *first/last*. Positive numbers count from the beginning of the dataset. Negative numbers count from the end of the dataset. Here are some examples:

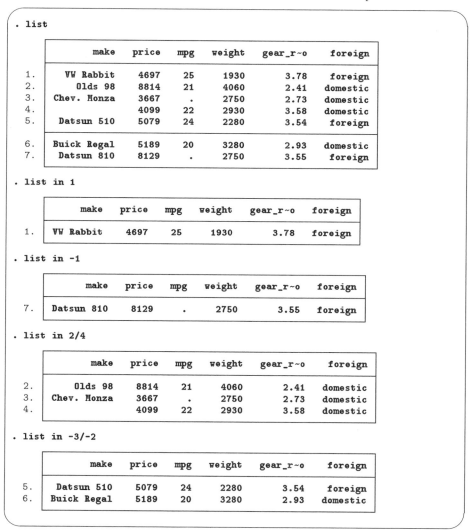

```
. list

              make    price   mpg   weight   gear_r~o   foreign

    1.     VW Rabbit    4697    25    1930       3.78    foreign
    2.       Olds 98    8814    21    4060       2.41   domestic
    3.   Chev. Monza    3667     .    2750       2.73   domestic
    4.                  4099    22    2930       3.58   domestic
    5.     Datsun 510   5079    24    2280       3.54    foreign

    6.    Buick Regal   5189    20    3280       2.93   domestic
    7.    Datsun 810    8129     .    2750       3.55    foreign

. list in 1

              make    price   mpg   weight   gear_r~o   foreign

    1.   VW Rabbit     4697    25    1930       3.78    foreign

. list in -1

              make    price   mpg   weight   gear_r~o   foreign

    7.   Datsun 810    8129     .    2750       3.55    foreign

. list in 2/4

              make    price   mpg   weight   gear_r~o   foreign

    2.       Olds 98    8814    21    4060       2.41   domestic
    3.   Chev. Monza    3667     .    2750       2.73   domestic
    4.                  4099    22    2930       3.58   domestic

. list in -3/-2

              make    price   mpg   weight   gear_r~o   foreign

    5.   Datsun 510    5079    24    2280       3.54    foreign
    6.   Buick Regal   5189    20    3280       2.93   domestic
```

Controlling the list output

The fine control over list output is exercised by specifying one or more options. You can use sepby() to separate observations by variable. abbreviate() specifies the minimum number of characters to abbreviate a variable in the output. divider draws a vertical line between the variables in the list.

```
. sort foreign
. list ma p g f, sepby(foreign)
```

		make	price	gear_r~o	foreign
1.	Chev. Monza		3667	2.73	domestic
2.	Olds 98		8814	2.41	domestic
3.			4099	3.58	domestic
4.	Buick Regal		5189	2.93	domestic
5.	VW Rabbit		4697	3.78	foreign
6.	Datsun 510		5079	3.54	foreign
7.	Datsun 810		8129	3.55	foreign

```
. list make weight gear, abbreviate(10)
```

		make	weight	gear_ratio
1.	Chev. Monza		2750	2.73
2.	Olds 98		4060	2.41
3.			2930	3.58
4.	Buick Regal		3280	2.93
5.	VW Rabbit		1930	3.78
6.	Datsun 510		2280	3.54
7.	Datsun 810		2750	3.55

```
. list, divider
```

		make	price	mpg	weight	gear_r~o	foreign
1.	Chev. Monza		3667	.	2750	2.73	domestic
2.	Olds 98		8814	21	4060	2.41	domestic
3.			4099	22	2930	3.58	domestic
4.	Buick Regal		5189	20	3280	2.93	domestic
5.	VW Rabbit		4697	25	1930	3.78	foreign
6.	Datsun 510		5079	24	2280	3.54	foreign
7.	Datsun 810		8129	.	2750	3.55	foreign

The separator() option draws a horizontal line at specified intervals. When not specified, it defaults to a value of 5.

```
. list, separator(3)

                   make     price     mpg    weight   gear_r~o    foreign

       1.    Chev. Monza     3667       .      2750       2.73    domestic
       2.        Olds 98     8814      21      4060       2.41    domestic
       3.                    4099      22      2930       3.58    domestic

       4.    Buick Regal     5189      20      3280       2.93    domestic
       5.      VW Rabbit     4697      25      1930       3.78     foreign
       6.     Datsun 510     5079      24      2280       3.54     foreign

       7.     Datsun 810     8129       .      2750       3.55     foreign
```

More

When you see a —more— prompt at the bottom of the Results window, it means that there is more information to be displayed. This happens, for example, when you are listing many observations.

```
. list make mpg

                   make      mpg

       1.    Linc. Continental    12
       2.    Linc. Mark V         12
       3.    Cad. Deville         14
       4.    Cad. Eldorado        14
       5.    Linc. Versailles     14

       6.    Merc. Cougar         14
       7.    Merc. XR-7           14
       8.    Peugeot 604          14
       9.    Buick Electra        15
      10.    Merc. Marquis        15

      11.    Buick Riviera        16
      12.    Chev. Impala         16
      13.    Dodge Magnum         16
      14.    Olds Toronado        16
      15.    AMC Pacer            17

      16.    Audi 5000            17
      17.    Dodge St. Regis      17
      18.    Volvo 260            17
      19.    Buick LeSabre        18
      20.    Dodge Diplomat       18

—more—
```

If you want to see the next screen of text, you have a few options: press any key, such as the Spacebar; click on the **More** button, ⬇ ; or click on the blue —more— at the bottom of the Results window. To see just the next line of text, press *Enter*.

Break

If you want to interrupt a Stata command, click on the **Break** button, 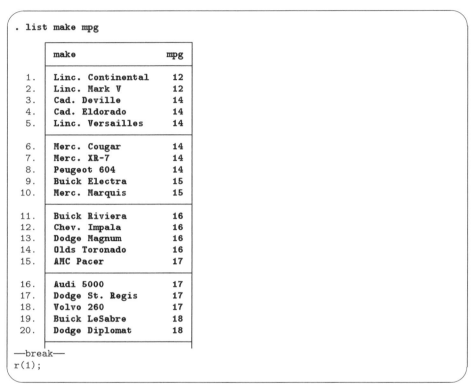. If you see a —more— prompt at the bottom of the Results window and wish to interrupt it, click on the **Break** button or press *q*.

```
. list make mpg

          |  make                mpg  |
          |-----------------------------|
      1.  |  Linc. Continental    12  |
      2.  |  Linc. Mark V         12  |
      3.  |  Cad. Deville         14  |
      4.  |  Cad. Eldorado        14  |
      5.  |  Linc. Versailles     14  |
          |-----------------------------|
      6.  |  Merc. Cougar         14  |
      7.  |  Merc. XR-7           14  |
      8.  |  Peugeot 604          14  |
      9.  |  Buick Electra        15  |
     10.  |  Merc. Marquis        15  |
          |-----------------------------|
     11.  |  Buick Riviera        16  |
     12.  |  Chev. Impala         16  |
     13.  |  Dodge Magnum         16  |
     14.  |  Olds Toronado        16  |
     15.  |  AMC Pacer            17  |
          |-----------------------------|
     16.  |  Audi 5000            17  |
     17.  |  Dodge St. Regis      17  |
     18.  |  Volvo 260            17  |
     19.  |  Buick LeSabre        18  |
     20.  |  Dodge Diplomat       18  |
          |-----------------------------|
—break—
r(1);
```

It is always safe to click on the **Break** button. After you click on **Break**, the state of the system is the same as if you had never issued the original command.

11 Creating new variables

generate and replace

This chapter shows the basics of creating and modifying variables in Stata. We saw how to work with the Data Editor in [GSU] **6 Using the Data Editor**—this chapter shows how we would do this from the Command window. The two primary commands used for this are

- generate for creating new variables. It has a minimum abbreviation of g.
- replace for replacing the values of an existing variable. It may not be abbreviated, because it alters existing data and hence can be considered dangerous.

The most basic form for creating new variables is generate *newvar* = *exp*, where *exp* is any kind of *expression*. Both generate and replace can be used with if and in qualifiers, of course. An expression is a formula made up of constants, existing variables, operators, and functions. Some examples of expressions (using variables from the auto dataset) would be 2 + price, weight^2, or sqrt(gear_ratio).

The operators defined in Stata are given in the table below:

	Arithmetic		Logical		Relational (numeric and string)
+	addition	!	not	>	greater than
−	subtraction	|	or	<	less than
*	multiplication	&	and	>=	> or equal
/	division			<=	< or equal
^	power			==	equal
				!=	not equal
+	string concatenation				

Stata has many mathematical, statistical, string, date, time-series, and programming functions. See help functions for the basics, and see [D] **functions** for a complete list and full details of all the built-in functions.

You can use menus and dialog boxes to create new variables and modify existing variables by selecting menu items from the **Data > Create or change data** menu. This feature can be handy for finding functions quickly. We will use the Command window for the examples in this chapter, however, because we would like to illustrate simple usage and some pitfalls.

generate

There are some details you should know about the generate command:

- The basic form of the generate command is generate *newvar* = *exp*, where *newvar* is a new variable name, and *exp* is any valid expression. You will get an error message if you try to generate a variable that already exists.
- An algebraic calculation using a missing value yields a missing value, as does division by zero, etc.

- If missing values are generated, the number of missing values in *newvar* is always reported. If Stata says nothing about missing values, then no missing values were generated.
- You can use generate to set the storage type of the new variable as it is generated. Saving an indicator variable as a byte can save size when used with a large dataset.

Here are some examples of creating new variables from the afewcars dataset that we have been using for the past few chapters. Some are nonsensical—they are only for illustrating the generate command. The last example shows a way to generate an indicator variable for cars weighing more than 3,000 pounds. Logical expressions in Stata result in "1" for "true" and "0" for "false." The if qualifier is needed to be sure that the computations are done only for observations where weight is not missing.

```
. list make mpg weight

            make    mpg   weight

  1.    VW Rabbit     25     1930
  2.       Olds 98    21     4060
  3.   Chev. Monza     .     2750
  4.                  22     2930
  5.    Datsun 510    24     2280

  6.   Buick Regal    20     3280
  7.    Datsun 810     .     2750

. generate lphk = 3.7854 * (100 / 1.6093) / mpg
(2 missing values generated)
. label var lphk "Liters per 100km"
. g strange = sqrt(mpg*weight)
(2 missing values generated)
. gen huge = weight >= 3000 if !missing(weight)
. l make mpg weight lphk strange huge

            make    mpg   weight      lphk     strange   huge

  1.    VW Rabbit     25     1930   9.408812   219.6588      0
  2.       Olds 98    21     4060  11.20097    291.9932      1
  3.   Chev. Monza     .     2750         .           .      0
  4.                  22     2930  10.69183    253.8897      0
  5.    Datsun 510    24     2280   9.800845   233.9231      0

  6.   Buick Regal    20     3280  11.76101    256.125       1
  7.    Datsun 810     .     2750         .           .      0
```

replace

Whereas `generate` is used to create new variables, `replace` is the command used for existing variables. Stata uses two different commands to prevent you from accidentally modifying your data. The `replace` command cannot be abbreviated. Stata generally requires you to spell out completely any command that can alter your existing data.

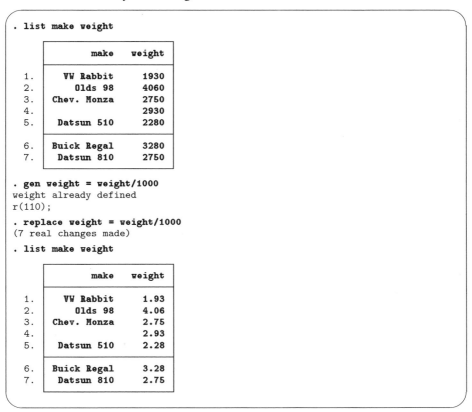

```
. list make weight

           make    weight

  1.    VW Rabbit     1930
  2.      Olds 98     4060
  3.  Chev. Monza     2750
  4.                  2930
  5.   Datsun 510     2280

  6.  Buick Regal     3280
  7.   Datsun 810     2750

. gen weight = weight/1000
weight already defined
r(110);

. replace weight = weight/1000
(7 real changes made)

. list make weight

           make    weight

  1.    VW Rabbit     1.93
  2.      Olds 98     4.06
  3.  Chev. Monza     2.75
  4.                  2.93
  5.   Datsun 510     2.28

  6.  Buick Regal     3.28
  7.   Datsun 810     2.75
```

Suppose that you want to create a new variable, `predprice`, which will be the predicted price of the cars in the following year. You estimate that domestic cars will increase in price by 5% and foreign cars, by 10%.

One way to create the variable would be to first use `generate` to compute the predicted domestic car prices. Then use `replace` to change the missing values for the foreign cars to their proper values.

(Continued on next page)

```
. gen predprice = 1.05*price if foreign==0
(3 missing values generated)
. replace predprice = 1.10*price if foreign==1
(3 real changes made)
. list make foreign price predprice, nolabel

            make   foreign   price   predpr~e
  1.    VW Rabbit        1    4697     5166.7
  2.      Olds 98        0    8814     9254.7
  3.   Chev. Monza       0    3667    3850.35
  4.                     0    4099    4303.95
  5.   Datsun 510        1    5079     5586.9

  6.   Buick Regal       0    5189    5448.45
  7.   Datsun 810        1    8129     8941.9
```

Of course, because foreign is an indicator variable, we could generate the predicted variable with one command:

```
. gen predprice2 = (1.05 + 0.05*foreign)*price
. list make foreign price predprice predprice2, nolabel

            make   foreign   price   predpr~e   predpr~2
  1.    VW Rabbit        1    4697     5166.7     5166.7
  2.      Olds 98        0    8814     9254.7     9254.7
  3.   Chev. Monza       0    3667    3850.35    3850.35
  4.                     0    4099    4303.95    4303.95
  5.   Datsun 510        1    5079     5586.9     5586.9

  6.   Buick Regal       0    5189    5448.45    5448.45
  7.   Datsun 810        1    8129     8941.9     8941.9
```

generate with string variables

Stata is smart. When you generate a variable and the expression evaluates to a string, Stata creates a string variable with a storage type as long as necessary, and no longer than that. `where` is a `str1` in the following example.

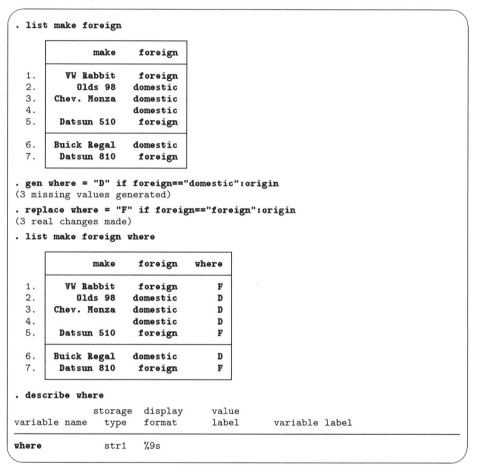

Stata has some useful tools for working with string variables. Here we split the `make` variable into make and model, and then create a variable that has the model together with where the model was manufactured.

(Continued on next page)

```
. gen model = substr(make, strpos(make," ")+1, .)
(1 missing value generated)

. gen modelwhere = model + " " + where

. list make where model modelwhere
```

	make	where	model	modelw~e
1.	VW Rabbit	F	Rabbit	Rabbit F
2.	Olds 98	D	98	98 D
3.	Chev. Monza	D	Monza	Monza D
4.		D		D
5.	Datsun 510	F	510	510 F
6.	Buick Regal	D	Regal	Regal D
7.	Datsun 810	F	810	810 F

There are a few things to note about how these commands worked:

1. $\text{strpos}(s_1, s_2)$ produces an integer equal to the first position in s_1 at which s_2 is found, or 0 if it is not found.

2. $\text{substr}(s, c_0, c_1)$ produces a string of length c_1 beginning with the c_0th character of s. If $c_1 = .$, it gives a string from character c_0 to the end of string.

3. Putting 1 and 2 together: $\text{substr}(s, \text{strpos}(s, " ")+1, .)$ produces s with its first word removed. Since make contains both the make and model of each car, and make contains no spaces in this dataset, we have found each car's model.

4. The operator '+', when applied to string variables, will concatenate the strings (i.e., join them together). The expression "this" + "that" results in the string "thisthat". When the variable modelwhere was generated, a space (" ") was added between the two strings.

5. The missing value for a string is nothing special—it is simply the empty string "". Thus the value of modelwhere for the car with no make or model is " D" (note the leading space).

12 Deleting variables and observations

clear, drop, and keep

In this chapter, we will present the tools for paring observations and variables from a dataset. We saw how to do this using the Data Editor in [GSU] **6 Using the Data Editor**; this chapter presents the methods for doing so from the Command window.

There are three main commands for removing data and other Stata objects from memory: `clear`, `drop`, and `keep`. Remember that they affect only what is in memory. None of these commands alters anything that has been saved to disk.

clear and drop _all

Suppose that you are working on an analysis or a simulation and you need to clear out Stata's memory so that you can impute different values or simulate a new dataset. You are not interested in saving any of the changes you have made to the dataset in memory—you would just like to have an empty dataset. What you do depends on how much you want to clear out: at any time, you can have not only data but also metadata such as value labels, saved results from previous commands, and saved matrices. The `clear` command will let you carefully clear out data or other objects; we are interested only in simple usage here. For more information, see `help clear` and [D] **clear**.

If you type the command `clear` into the Command window, it will remove all variables and value labels. In basic usage, this is typically enough. It has the nice property that it does not remove any saved results, so you can load a new dataset and predict values by using saved estimation results from a model fit on a previous dataset. See `help postest` and [U] **20 Estimation and postestimation commands** for more information.

If you want to be sure that everything is cleared out, use the command `clear all`. This will clear Stata's memory of data and all auxiliary objects so that you can start with a clean slate. The first time you use `clear all` while you have a graph or dialog open, you may be surprised when that graph or dialog closes; this is necessary so that Stata can free all memory that is being used.

If you want to get rid of just the data and nothing else, you can use the command `drop _all`.

drop

The `drop` command is used to remove variables or observations from the dataset in memory.
- If you want to drop variables, use `drop` *varlist*.
- If you want to drop observations, use `drop` with an `if` or an `in` qualifier, or both.

(Continued on next page)

We will use the `afewcars` dataset, as usual.

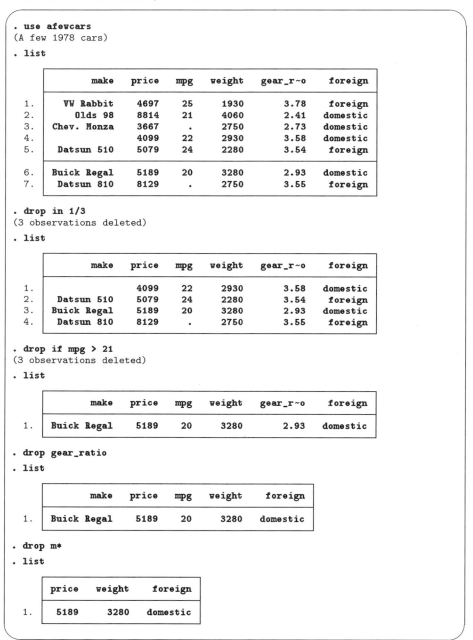

```
. use afewcars
(A few 1978 cars)
. list
```

	make	price	mpg	weight	gear_r~o	foreign
1.	VW Rabbit	4697	25	1930	3.78	foreign
2.	Olds 98	8814	21	4060	2.41	domestic
3.	Chev. Monza	3667	.	2750	2.73	domestic
4.		4099	22	2930	3.58	domestic
5.	Datsun 510	5079	24	2280	3.54	foreign
6.	Buick Regal	5189	20	3280	2.93	domestic
7.	Datsun 810	8129	.	2750	3.55	foreign

```
. drop in 1/3
(3 observations deleted)
. list
```

	make	price	mpg	weight	gear_r~o	foreign
1.		4099	22	2930	3.58	domestic
2.	Datsun 510	5079	24	2280	3.54	foreign
3.	Buick Regal	5189	20	3280	2.93	domestic
4.	Datsun 810	8129	.	2750	3.55	foreign

```
. drop if mpg > 21
(3 observations deleted)
. list
```

	make	price	mpg	weight	gear_r~o	foreign
1.	Buick Regal	5189	20	3280	2.93	domestic

```
. drop gear_ratio
. list
```

	make	price	mpg	weight	foreign
1.	Buick Regal	5189	20	3280	domestic

```
. drop m*
. list
```

	price	weight	foreign
1.	5189	3280	domestic

These changes are only to the data in memory. If you wanted to make the changes permanent, you would need to save the dataset.

keep

keep tells Stata to drop all variables except those specified explicitly or through the use of an `if` or `in` expression. Just like `drop`, `keep` can be used with a *varlist* or with qualifiers, but not with both at once. We use a `clear` command at the start of this example, so that we can reload the `afewcars` dataset.

```
. clear

. use afewcars
(A few 1978 cars)

. list
```

	make	price	mpg	weight	gear_r~o	foreign
1.	VW Rabbit	4697	25	1930	3.78	foreign
2.	Olds 98	8814	21	4060	2.41	domestic
3.	Chev. Monza	3667	.	2750	2.73	domestic
4.		4099	22	2930	3.58	domestic
5.	Datsun 510	5079	24	2280	3.54	foreign
6.	Buick Regal	5189	20	3280	2.93	domestic
7.	Datsun 810	8129	.	2750	3.55	foreign

```
. keep in 4/7
(3 observations deleted)

. list
```

	make	price	mpg	weight	gear_r~o	foreign
1.		4099	22	2930	3.58	domestic
2.	Datsun 510	5079	24	2280	3.54	foreign
3.	Buick Regal	5189	20	3280	2.93	domestic
4.	Datsun 810	8129	.	2750	3.55	foreign

```
. keep if mpg <= 21
(3 observations deleted)

. list
```

	make	price	mpg	weight	gear_r~o	foreign
1.	Buick Regal	5189	20	3280	2.93	domestic

```
. keep m*

. list
```

	make	mpg
1.	Buick Regal	20

Notes

13 Using the Do-file Editor—automating Stata

The Do-file Editor in Stata(GUI)

Stata comes with an integrated text editor called the Do-file Editor, which can be used for many tasks. It gets its name from the term *do-file*, which is a file containing a list of commands for Stata to run (what is called a batch file in other settings). See [U] **16 Do-files** for more information. Although the Do-file Editor has advanced features that can help in writing such files, it can also be used to build up a series of commands that can then be submitted to Stata all at once. This feature can be handy when writing a loop to process multiple variables in a similar fashion or when doing complex repetitive tasks interactively.

To get the most from this chapter, you should work through it at your computer.

The Do-file Editor toolbar

The Do-file Editor has 13 buttons. Many of the buttons have a similar purpose as their look-alikes in the main Stata toolbar.

If you ever forget what a button does, hold the mouse pointer over a button for a moment, and a tooltip will appear with a description of that button.

New: Open a new do-file in a new Do-file Editor.

Open: Open a do-file from disk in a new Do-file Editor.

Save: Save the current do-file to disk.

Print: Print the contents of the Do-file Editor.

Find: Open the *Find/Replace* dialog for finding and replacing text.

Cut: Cut the selected text and put it in the Clipboard.

Copy: Copy the selected text to the Clipboard.

Paste: Paste the text from the Clipboard into the current document.

Undo: Undo the last change.

Redo: Undo the last undo.

Show File in Viewer: Show the contents of the do-file in a Viewer window. This is worthwhile when editing files containing SMCL tags, such as log files or help files.

Execute Quietly (run): Run the commands in the do-file without showing any output. We will refer to this as the **Run** button.

Execute (do): Run the commands in the do-file, showing all commands and their output. We will refer to this as the **Do** button.

Using the Do-file Editor

Suppose that we would like to analyze fuel usage for 1978 automobiles similarly to what we did in [GSU] **1 Introducing Stata—sample session**. We know that we will be issuing many commands to Stata during our analysis and that we want to be able to reproduce our work later without having to type each command again.

We can do this easily in Stata: simply save a text file containing the commands. When this is done, we can tell Stata to run the file and execute each command in sequence. Such a file is known as a Stata *do-file*; see [U] **16 Do-files**.

To analyze fuel usage of 1978 automobiles, we would like to create a new variable giving gallons per mile. We would like to see how that variable changes in relation to vehicle weight for both domestic and imported cars. Performing a regression with our new variable would be a good first step.

To get started, click on the **Do-file Editor** button to open the Do-file Editor. After the Do-file Editor opens, type the commands below, because they are what we would like to submit. Purposely misspell the name of the `foreign` variable on the fifth line. (We are intentionally making some common mistakes and then pointing you to the solutions. This will save you time later.)

```
* an example do-file
sysuse auto
generate gpm = 1/mpg
label var gpm "Gallons per mile"
regress gpm weight foreing
```

Here is what your Do-file Editor should look like now:

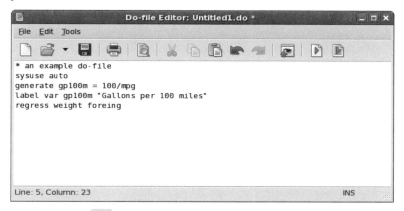

Click on the **Do** button, , to run the commands: When you click on the **Do** button, Stata executes the commands in sequence, and the results appear in the Results window:

```
. do /tmp/SD00001.000000
. * an example do-file
. sysuse auto
(1978 Automobile Data)
. generate gpm = 1/mpg
. label var gpm "Gallons per mile"
. regress gpm weight foreing
variable foreing not found
r(111);
.
end of do-file
```

The do "/tmp/..." command is how Stata executes the commands in the Do-file Editor. Stata saves the commands to a temporary file and issues the do command to execute them.

(Continued on next page)

Everything worked as planned until Stata saw the misspelled variable. The first three commands were executed, but an error was produced on the fourth. Stata does not know of a variable named `foreing`. We need to go back to the Do-file Editor and correct the mistake:

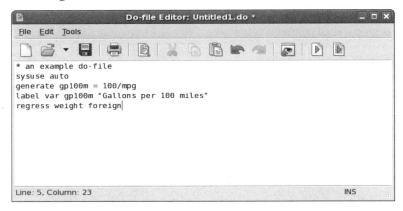

We click on the Do button again. Alas, Stata now fails on the first line—it will not overwrite the dataset in memory that we had changed.

```
. do /tmp/SD00001.000000
. * an example do-file
. sysuse auto
no; data in memory would be lost
r(4);
```

We now have a choice for what we should do:

- We could use our do-file to automatically clear out Stata's memory before it runs. Doing so is convenient, but dangerous, because it defeats Stata's protection against throwing away changes without warning.
- We could manually clear the dataset and then process the do-file again. This process can be aggravating when building a complicated do-file.

Here is some advice: Automatically clear Stata's memory while debugging the do-file. Once the do-file is in its final form, decide the context in which it will be used. If it will be used in a highly automated environment (such as when certifying), the do-file should still automatically clear Stata's memory. If it will be used rarely, do not clear Stata's memory. It will save heartache.

We will add a `clear` option to the `sysuse` command to automatically clear the dataset in Stata's memory before the do-file runs:

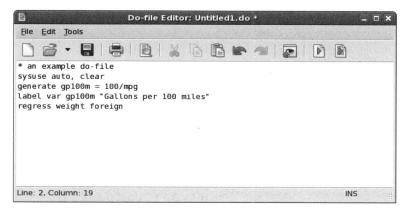

The do-file now runs well, as clicking on the Do button shows:

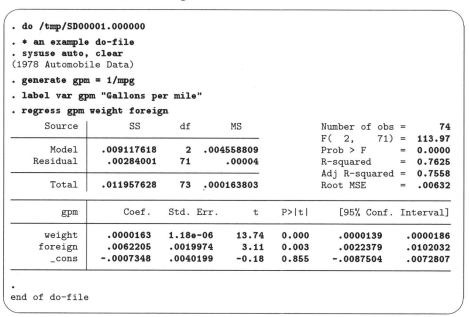

```
. do /tmp/SD00001.000000

. * an example do-file

. sysuse auto, clear
(1978 Automobile Data)

. generate gpm = 1/mpg

. label var gpm "Gallons per mile"

. regress gpm weight foreign

      Source |       SS       df       MS              Number of obs =      74
-------------+------------------------------           F(  2,    71) =  113.97
       Model |  .009117618     2   .004558809           Prob > F      =  0.0000
    Residual |   .00284001    71      .00004           R-squared     =  0.7625
-------------+------------------------------           Adj R-squared =  0.7558
       Total |  .011957628    73   .000163803           Root MSE      =  .00632

-------------+----------------------------------------------------------------
         gpm |      Coef.   Std. Err.      t    P>|t|     [95% Conf. Interval]
-------------+----------------------------------------------------------------
      weight |   .0000163   1.18e-06    13.74   0.000     .0000139    .0000186
     foreign |   .0062205   .0019974     3.11   0.003     .0022379    .0102032
       _cons |  -.0007348   .0040199    -0.18   0.855    -.0087504    .0072807
-------------+----------------------------------------------------------------

.
end of do-file
```

You might want to select **File > Save As...** from the Do-file Editor to save this do-file. Later, you could select **File > Open...** to open it and then add more commands as you move forward with your analysis. By saving the commands of your analysis in a do-file as you go, you do not have to worry about retyping them with each new Stata session. Think hard about removing the `clear` option from the first command.

After you have saved your do-file, you can execute the commands it contains by typing do *filename*, where the *filename* is the name of your do-file.

The File menu

The **File** menu of the Do-file Editor includes standard features found in most text editors. You may choose to start a **New** file, **Open...** an existing file, **Save** the current file, or save the current file under a new name with **Save As...**. You may also **Print...** the current file. There are also buttons on the Do-file Editor's toolbar that correspond to these features.

You can also select **File > Insert File...** to insert the contents of another file at the current cursor position in the Do-file Editor.

Editing tools

The **Edit** menu of the Do-file Editor includes the standard **Undo**, **Redo**, **Cut**, **Copy**, and **Paste** capabilities. There are also buttons on the Do-file Editor's toolbar for easy access to these capabilities. There are several other **Edit** menu features that you may find useful:

- You may delete or select the current line.
- You may also shift the current line or selection right or left one tab stop.
- **Make Selection Uppercase** converts the selection to all capital letters.
- **Make Selection Lowercase** converts the selection to all lowercase letters.

Searching

Stata's Do-file Editor includes standard **Find...** and **Replace...** capabilities in the **Edit > Find** menu. You should already be familiar with these capabilities from other text editors and word processors. You may access the **Find** capability from a button on the toolbar as well.

The **Edit > Find** menu also includes some choices that are useful when you are writing a long do-file or ado-file. You may select **Edit > Find > Go to Line...** and jump straight to any line in your file.

Matching and balancing of parentheses ( ), braces { }, and brackets [ ] are available from the **Edit > Find** menu. When you select **Edit > Find > Match Brace**, the Do-file Editor looks at the character immediately to the right of the cursor. If it is one of the characters that the editor can match, the editor will find the matching character and place the cursor immediately in front of it. If there is no match, you will hear a beep and the cursor will not move.

When you select **Edit > Find > Balance Braces**, the Do-file Editor looks to the left and right of the current cursor position or selection and creates a selection including the narrowest level of matching characters. If you select **Balance Braces** again, the editor will expand the selection to include the next level of matching characters. If there is no match, you will hear a beep and the cursor will not move.

Balance Braces is easier to explain with an example. Type `(now (is the) time)` in the Do-file Editor. Place the cursor between the words `is` and `the`. Select **Edit > Find > Balance Braces**. The Do-file Editor will select `(is the)`. If you select **Balance Braces** again, the Do-file Editor will select `(now (is the) time)`.

The Tools menu

You have already learned about the **Do** button. Selecting **Tools > Execute (do)** is equivalent to clicking on the **Execute (do)** button.

You have also already learned about the **Run** button. Selecting **Tools > Execute Quietly (run)** is equivalent to clicking on the **Execute Quietly (run)** button. Executing quietly still executes all the commands in the Do-file Editor, but does not produce any output. It is unlikely that you will ever need to use the **Run** button; see [U] **16.4.2 Suppressing output** for more details.

Do and **Run** are equivalent to Stata's do and run commands; see [U] **16 Do-files** for a complete discussion.

If you wish to execute a subset of the lines in your do-file, highlight those lines, and select **Tools > Execute (do)**. If you wish to execute all commands from the current line through the end of the file, select **Tools > Execute (do) to Bottom**.

You can also preview files in the Viewer by selecting **Tools > Show File in Viewer** or by clicking on the **Show File in Viewer** button, . This feature is useful when working with files that use Stata's SMCL tags, such as when writing help files or editing log files.

Saving interactive commands from Stata as a do-file

While working interactively with Stata, you may decide that you would like to rerun the last several commands that you typed interactively. From the Review window, you can send highlighted commands or even the entire contents to the Do-file Editor. You can also save commands as a do-file and open that file in the Do-file Editor. You can copy a command from a dialog (rather than submit it) and paste it into the Do-file Editor. See [GSU] **6 Using the Data Editor** for details. Also see [R] **log** for information on the cmdlog command, which allows you to log all commands that you type in Stata to a do-file.

Notes

14 Graphing data

Working with graphs

Stata has a rich system for graphical representation of data. The main command for creating graphs is unsurprisingly named graph. Behind this plain name is a wealth of tools. In this chapter, we will make one simple graph to point out the basics of the Graph window. See the *Graphics Reference Manual* for more information about all aspects of working with graphs.

A simple graph example

In the sample session of [GSU] **1 Introducing Stata—sample session**, we made a scatterplot, added a fitted regression line, and made a grid of scatterplots to allow comparisons across groups. Here we make a simple box plot that shows the displacements of the cars' engines and how they compare across repair records within the place of manufacture of the cars, using the auto dataset (sysuse auto).

We select **Graphics > Box plot**, choose displacement in the *Variables* field on the **Main** tab, click on the **Categories** tab, check the *Group 1* checkbox and enter rep78 for the first group, and check the *Group 2* checkbox and enter foreign for the second group. Finally, we click on the **Submit** button so that we could easily make changes to the graph if need be. After we look at the graph, we realize that we forgot the title. We close the Graph window, click on the **Titles** tab of the *graph box* dialog, add the title *Displacement across Repairs within Origin*, and click on the **Submit** button again.

The Graph window comes up, showing us the fruit of our labor:

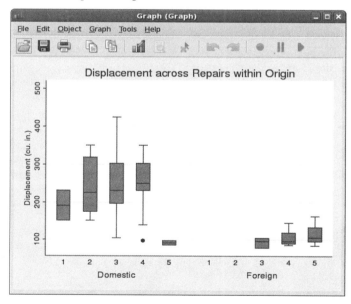

Graph window

When the Graph window comes up, it shows our graph in a window with a toolbar. The first four icons are familiar to us from other Stata windows: **Open**, **Save**, **Print**, and **Copy**. The next two icons are new:

 Rename Graph: This allows the graph to be renamed. Why would you do this? If you would like to have multiple graphs open at once, the graphs need to be named. So, you can click on the **Rename** button to give a graph a name. This window will then remain open when you create your next graph.

 Graph Editor: Stata has a Graph Editor that allows you to manipulate and edit your graph. This feature will be introduced in the next chapter.

The inactive buttons to the right of the **Graph Editor** button are used by the Graph Editor, so their meanings will become clear in the next chapter.

We decide that we like this graph and would like to save it. We can save it either by clicking on the **Save** button and choosing a name and a location or by right-clicking on the Graph window itself, and selecting **Save As...**.

Saving and printing graphs

You can save a graph once it is displayed by right-clicking on its window and selecting **Save As...**. You can print a graph by right-clicking on its window and selecting **Print...**. You can also use the **File** menu to save or print a graph. We recommend that you always right-click on a graph to save or print it to ensure that the correct graph is selected.

Right-clicking on the Graph window

Right-clicking on the Graph window displays a menu from which you can select the following:
- **Save As...** to save the graph to disk.
- **Copy** to copy the graph to the Clipboard.
- **Start Graph Editor** to start the Graph Editor.
- **Preferences...** to edit the preferences for graphs.
- **Print...** to print the contents of the Graph window.

The Graph button

The **Graph** button, , is located on the main window's toolbar. The button has two parts, an icon and an arrow. Clicking on the icon brings the topmost Graph window to the front of all other windows. Clicking on the arrow displays a menu of open graphs. Selecting a graph from the menu brings that graph to the front of all other windows. If you ever decide to close the Graph window, you can reopen it only by reissuing a Stata command that draws a new graph.

15 Editing graphs

Working with the Graph Editor

With Stata's Graph Editor, you can change almost anything on your graph, and you can add text, lines, arrows, and markers wherever you would like.

We will first make a graph to edit and will then point out the tools in the Graph Editor. Here is the command that we'll use to make the graph:

```
. scatter mpg weight, name(mygraph) title(Mileage vs. Vehicle Weight)
```

Start the Editor by right-clicking on your graph and selecting **Start Graph Editor**. Click once on the title of the graph. Here is a picture of the Graph Editor with its elements labeled.

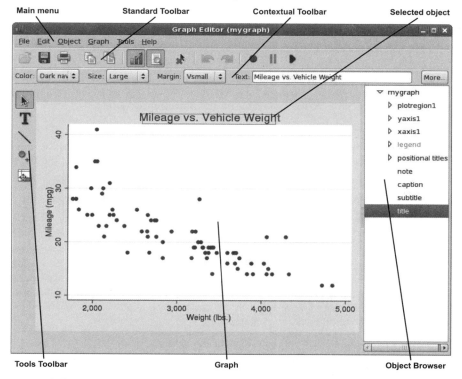

Select any of the tools along the left of the Editor to edit the graph. The Pointer (Select Tool), 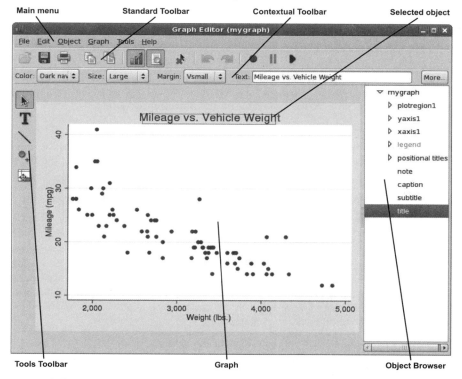, is selected by default.

You can change the properties of objects or drag them to new locations by using the Pointer. As you select objects with the Pointer, a Contextual Toolbar will appear just above the graph. In our example, the title of the graph is selected, so the Contextual Toolbar has controls that are relevant for editing titles. You can use any of the controls on the Contextual Toolbar to immediately change the most important properties of the selected object. Right-click on an object to access more properties and operations. Hold the *Shift* key when dragging objects to constrain the movement to horizontal, vertical, or 90-degree angles.

Add text, lines, or markers (with optional labels) to your graph by using the three *Add...* tools— **T** , **** , and ○+ . Lines can be changed to arrows by using the Contextual Toolbar. If you do not like the default properties, simply change their settings in the Contextual Toolbar before adding the text, line, or marker. The new setting will then be applied to all added objects, even in future Stata sessions.

Do not be afraid to try things. If you do not like a result, change it back by using the same tool, or click on the **Undo** button, ↰, in the Standard Toolbar for the Graph Editor (below the main menu). **Edit > Undo** in the main menu does the same thing.

Remember to reselect the Pointer tool when you want to drag objects or change their properties.

You can move objects on the graph and have the rest of the objects adjust their position to accommodate the move with the **Grid Edit** tool, ⊞ . With this tool, you are repositioning objects in the underlying grid that holds the objects in the graph. Some graphs, e.g., by graphs, are composed of nested grids. You can reposition objects only within the grid that contains them; they cannot be moved to other grids.

You can also select objects in the Object Browser along the right of the graph. This window shows a hierarchical listing of the objects in the graph. Clicking or right-clicking on an object in the Object Browser is the same as clicking or right-clicking on the object in the graph.

The Graph Editor has the ability to record your actions and play them back on later graphs. When you click on the **Record** button, ● , every editing action you take, including undos and redos, is recorded. If you would like to do some editing that is not recorded, you can click on the **Pause** button, **||** . You can click on the **Pause** button again to resume recording. When you are done with your recording, click on the **Record** button. You will be prompted to save your recording. Any recording you save is available from the **Play Recording** button, ▶ , and may be applied to future graphs. You can even play a recording in any Stata graph command by using the play option. See *Graph Recorder* in [G] **graph editor** for more information.

Stop the editor by selecting **File > Stop Graph Editor** from the main menu or by clicking on the **Graph Editor** button. When you stop the Graph Editor, you will be prompted to save your graph if you have made any changes. If you do not save your graph, your changes will not be lost, but you will risk losing them if you create a new graph in the same Graph window. You must stop the Editor if you would like to work on other tasks in Stata.

Here are a few of the things that you can do with the Editor:
- Add annotations using lines, arrows, and text.
- Add or remove grid lines or reference lines.
- Add or modify titles, captions, and notes.
- Change scatterplots to line plots, connected plots, areas, bars, spikes, or drop lines—and, of course, vice versa.
- Change the size, color, margin, and other properties of your graph's titles (or any other text on the graph).
- Move your legend to another side of the graph, or even place it in the plot region.
- Change the aspect ratio of your graph.
- Stack the bars on a bar graph or turn them into percentages.

- Rotate or change the angle of axis labels.
- Add custom ticks and labels to the axes.
- Change the rule for the number and spacing of ticks and labels on an axis.
- Emphasize a point on the graph, whether marker, bar, spike, or other plot, by making it a custom color, size, or symbol.
- Change the text or properties of a marker label.

Because you can edit every property of every object on the graph, you can change almost anything about your graph. To learn more, see [G] **graph editor** or type help graph editor.

Notes

16 Saving and printing results by using logs

Using logs in Stata

When working on an analysis, it is worthwhile to behave like a bench scientist and keep a lab notebook of your actions so that your work can be easily replicated. Everyone has a feeling of complete omniscience while working intensely—this feeling is wonderful but fleeting. The next day, the exact small details needed for perfect duplication have become obscure. Stata has a lab notebook at hand: the *log* file.

A log file is simply a record of your Results window. It records all commands and all textual output as it happens. Thus it keeps your lab notebook for you as you work. Because it writes the file while it writes the Results window, it also protects you from disastrous failures, be they power failures or computer crashes. We recommend that you start a log file whenever you begin any serious work in Stata.

Logging output

All the output that appears in the Results window can be captured in a log file. Stata can save the file in one of two different formats. By default, Stata will save the file in its Stata Markup and Control Language (SMCL) format, which preserves all the formatting and links from the Results window. You can open these results in the Viewer, and they will behave as though they were in the Results window. If you would rather have plain-text files without any formatting, you can save the file as a plain log file. We recommend using the SMCL format, because SMCL files can be translated into a variety of formats readable by applications other than Stata with the **File > Log > Translate...** menu (see [R] **translate**).

To start a log file, click on the **Log** button, . This will open a standard file dialog that allows you to specify a directory and filename for your log. If you do not specify a file extension, the extension .smcl will be added to the filename. If you specify a file that already exists, you will be asked whether you want to append the new log to the file or overwrite the file with the new log.

(Continued on next page)

111

Here is an example of a short session:

```
        name:  <unnamed>
         log:  /home/mydir/data/base.smcl
    log type:  smcl
   opened on:  11 May 2009, 22:26:49
. sysuse auto
(1978 Automobile Data)
. by foreign, sort: summarize price mpg

-> foreign = Domestic
    Variable |       Obs        Mean    Std. Dev.       Min        Max
       price |        52    6072.423    3097.104      3291      15906
         mpg |        52    19.82692    4.743297        12         34

-> foreign = Foreign
    Variable |       Obs        Mean    Std. Dev.       Min        Max
       price |        22    6384.682    2621.915      3748      12990
         mpg |        22    24.77273    6.611187        14         41
. * be sure to include the above stats in report!
. * now for something completely different
. corr price mpg
(obs=74)
             |    price      mpg
       price |   1.0000
         mpg |  -0.4686   1.0000
. log close
        name:  <unnamed>
         log:  /home/mydir/data/base.smcl
    log type:  smcl
   closed on:  11 May 2009, 22:26:50
```

There are a few items of interest.

- The header showing the log file's location, type, and starting timestamp is part of the log file. This feature helps when working with multiple log files.
- The two lines starting with asterisks (*) are *comments*. Stata ignores the text following the asterisk, so you may type any comment you would like, with any special characters you would like. Commenting is a good way to document your thought process and to mark sections of the log for later use.
- In this example, the log file was closed using the log close command. Doing so is not strictly necessary because log files are automatically closed when you exit Stata.

Stata allows multiple log files to be open at once only if the log files are *named*. For details on this, see help log.

Working with logs

Log files are best viewed using Stata's Viewer. Select **File > Log > View...**. If there is a log file open (as shown by the status bar), it will be the default log file to view; otherwise, you need to either type the name of the log file into the dialog or click on the **Browse...** button to find the file with a standard file dialog.

Once you are in the Viewer window, everything behaves as expected: you can copy text and paste between the Viewer and anything else that uses text, such as word processors or text editors. You can even paste into the Command window or the Do-file Editor, but you should take care to copy only commands, not their output. It is OK to copy the prompt (".") at the start of the echoed command, because Stata is smart enough to ignore it in the Command window. When working with a word processor, what you paste will be unformatted text; it will look best if you use a fixed-width font, like Courier, to display it.

Viewing your current log file is a good way to keep a reminder of something you have already done or a view of a previous result. The Viewer window takes a snapshot of your log file and hence will not scroll as you keep working in Stata. If you need to see more recent results in the Viewer, press the **Refresh** button.

For more detailed information about logs, see [U] **15 Saving and printing output—log files** and [R] **log**. For more information about the Viewer, see [GSU] **3 Using the Viewer**.

Printing logs

To print a standard SMCL log file, you need to have the log file open in a Viewer window. Once the log file is in the Viewer, you can either right-click on the Viewer window, and select **Print...**, or you can select the Viewer window from **File > Print >** *"Viewer title"*. A *Print* dialog will appear.

- You can fill in none, any, or all of the items *Header, Name,* and *Project*. You can check or uncheck options to *Print Line #s, Print Header,* and *Print Logo*. These items are saved and will appear again in the *Print* dialog (in this and in future Stata sessions).

- Stata, by default, will choose a font size that it thinks is appropriate for your printer.

You could also use the `translate` command to generate a Postscript file. See [R] **translate** for more information.

If your log file is a plain-text file (`.log` instead of `.smcl`), you can load it into a text editor, such as Emacs or vi, the Do-file Editor in Stata, or your favorite word processor. You can then edit the log file—add headings, comments, etc.—format it, and print it. If you bring the log file into a word processor, it will be displayed and printed with its default font. The log file will not be easily readable when printed in a proportionally spaced font (e.g., Times Roman or Helvetica). It will look much better printed in a fixed-width font (e.g., Courier).

Rerunning commands as do-files

Stata can also log just the commands from a session without recording the output. This feature is a convenient way to make a do-file interactively. Such a file is called a *cmdlog* file by Stata. You can start a `cmdlog` file by typing

    cmdlog using *filename*

and you can close the `cmdlog` file by typing

    cmdlog close

Here, for example, is what a `cmdlog` of the previous session would look like. It contains only commands and hence could be used as a do-file.

```
sysuse auto
by foreign, sort: summarize price mpg
* be sure to include the above stats in report!
* now for something completely different
corr price mpg
```

If you start working and then wish you had started a `cmdlog` file, you can save yourself heartache by saving the contents of the Review window. The Review window stores the last 5,000 commands you have typed. Simply right-click on the Review window and select **Save All...** from the menu. If you would like to move the commands directly to the Do-file Editor, select **Send to Do-file Editor**. You may find this method a more convenient way to create a text file containing only the commands that you typed during your session.

See [GSU] **13 Using the Do-file Editor—automating Stata**, [U] **16 Do-files**, and [U] **15 Saving and printing output—log files** for more information.

17 Setting font and window preferences

Changing and saving fonts and sizes and positions of your windows

You may find that you would like to change the fonts and display style of Stata's windows, depending on your monitor resolution and personal preferences. At the same time, there could be requirements for font usage, say, when you submit graphs to journals. Stata accommodates both of these by allowing sets of preferences for how windows are displayed.

We will first cover what can be changed in each window and then talk about what you can manage with your preferences.

Graph window

The preferences for the Graph window can be changed by right-clicking on the Graph window and choosing **Preferences...** from the contextual menu. The settings can then be set for graphs as they display in Stata. The settings that should be used when printing can be set using the **Printer** item.

The Graph preferences allow various *schemes*. These provide a quick way to optimize graphs for printing or display on a screen. There are even schemes defined for *The Economist* and the *Stata Journal* so that you can get the details for these journals right without much fuss. Changing the scheme *does not* change the current graph—it applies the settings to future graphs.

All other windows

You can change the display font and font size for all other types of windows in Stata.

Fonts and font sizes can be changed for any type of window by right-clicking on the window and selecting **Preferences...**. Doing so will bring up the Preferences dialog, from which you can pick the font and size of your choice. You should take care to choose a fixed-width font for the Results, Viewer, Data Editor, and Do-file Editor windows. You are not prevented from choosing other fonts, but if you do choose a proportional-width font, output and numbers will not align as they should.

Changing color schemes

The Results and Viewer windows have color schemes that control the way in which input, text, results, errors, links, and highlighted text display. Each has its color scheme set in the same fashion: you can right-click on the window and select or design your own color scheme. The default setting for both the Results window and the Viewer is the built-in *White Background* scheme. There are six other built-in schemes and three settings for custom schemes. The settings for the Viewer affect all Viewer windows at once.

(Continued on next page)

Managing multiple sets of preferences

Stata's preferences are automatically saved when you exit Stata, and they are reloaded when Stata is launched. However, sometimes you may wish to rearrange Stata's windows and then revert the windows to your preferred arrangement. You can do this by saving your preferences to a file and loading them later. Any changes you make to Stata's preferences after loading a preferences file do not affect the file; the file remains untouched unless you specifically overwrite it.

To manage preferences, open the **Edit > Preferences > Manage Preferences** menu, and do any of the following:

- Select **Open Preferences...** to open a preferences file that is not in the default location in your Library folder.
- Select **Save Preferences...** to save the current window arrangement and preferences to disk. By default, these are saved in your .user_prefs folder in your home directory. If you save your preferences to this default folder, they will appear in the **Stata > Preferences > Manage Preferences** menu the next time you view it.
- Select **Factory Settings** to restore all preferences to their original settings.

Closing and opening windows

You can close the Viewer, Graph, Do-file Editor, and Data Editor windows. If you want to open a closed window, open the **Window** menu and select the desired window. You can cannot close the Command and Results windows.

18 Learning more about Stata

Where to go from here

You now know plenty enough to use Stata. There is still much, much more to learn, because Stata is a rich environment for doing statistical analysis and data management. What should you do to learn more?

- Get an interesting dataset and play with Stata.
  a. Use the menus and dialog system to experiment with commands. Notice what commands show up in the Results window. You will find that Stata's simple and consistent command syntax will make the commands easy to read so that you will know what you have done, and easy to remember so that typing some commands will be faster than using menus.
  b. Play with graphs and the Graph Editor.

- If you venture into the Command window, you will find that many things will go faster. You will also find that it is possible to make mistakes where you cannot understand why Stata is balking.
  a. Try `help` *commandname* or **Help > Stata Command...** and entering the command name.
  b. Look at the command syntax, and the examples in the help file, and compare them with what you typed. Compare them closely: small typos make commands impossible for Stata to parse.

- Explore Stata by selecting **Help > Search...** and choosing *Search documentation and FAQs*. You will uncover many statistical routines that could be of great use. Explore beyond this by using the `findit` command.

- Look through the *Combined subject table of contents* in the *Quick Reference and Index*.

- Read and work your way through the *User's Guide*. It is designed to be read cover to cover, and it contains most of the information you need to become an expert Stata user. It is well worth reading.

- Flip through the reference manuals, either paper or PDF, to read about statistical methods you like to use. The reference manuals are not meant to be read cover to cover—they are meant to be referred to like you would an encyclopedia. You can find the datasets used in the examples in the manuals by selecting **File > Example Datasets...**, and then clicking on `Stata 11 manual datasets`. Doing so will enable you to work through the examples quickly.

- Stata has much information, including answers to frequently asked questions (FAQs), at http://www.stata.com/support/faqs/.

- There are many useful links to Stata resources at http://www.stata.com/links/resources.html. Be sure to look at these materials, as many outstanding resources about Stata are listed here.

- Join Statalist, a listserver devoted to discussion of Stata and statistics.

- If you prefer to sample the *User's Guide* and the references, there is some advice later in this chapter for you.

- Subscribe to the *Stata Journal*, which contains reviewed papers, regular columns, book reviews, and other material of interest to researchers applying statistics in a variety of disciplines. Visit http://www.stata-journal.com.

- Many supplementary books about Stata are available. Visit the Stata bookstore at http://www.stata.com/bookstore/.

- Take a Stata NetCourse™. NetCourse 101 is an excellent choice to learn about Stata. See http://www.stata.com/netcourse/ for course information and schedules.

- Attend a public training course taught by StataCorp at third-party sites. Visit http://www.stata.com/training/public.html for course information and schedules.

Suggested reading from the User's Guide and reference manuals

The *User's Guide* is designed to be read from cover to cover. The reference manuals are designed as references to be sampled when necessary.

Ideally, after reading this *Getting Started* manual, you should read the *User's Guide* from cover to cover, but you probably want to become at least somewhat proficient in Stata right away. Here is a suggested reading list of sections from the *User's Guide* and the reference manuals to help you on your way to becoming a Stata expert.

This list covers fundamental features and points you to some less obvious features that you might otherwise overlook.

Basic elements of Stata
    [U] **11 Language syntax**
    [U] **12 Data**
    [U] **13 Functions and expressions**

Memory
    [U] **6 Setting the size of memory**
    [D] **compress** — Compress data in memory

Data input
    [U] **6 Setting the size of memory**
    [U] **21 Inputting data**
    [D] **infile** — Overview of reading data into Stata
    [D] **insheet** — Read ASCII (text) data created by a spreadsheet
    [D] **append** — Append datasets
    [D] **merge** — Merge datasets

Graphics
    Graphics Reference Manual

Useful features that you might overlook
    [U] **28 Using the Internet to keep up to date**
    [U] **16 Do-files**
    [U] **19 Immediate commands**
    [U] **23 Working with strings**
    [U] **24 Working with dates and times**
    [U] **25 Working with categorical data and factor variables**
    [U] **13.5 Accessing coefficients and standard errors**
    [U] **13.6 Accessing results from Stata commands**
    [U] **26 Overview of Stata estimation commands**
    [U] **20 Estimation and postestimation commands**
    [R] **estimates** — Save and manipulate estimation results

Basic statistics

[R] **anova** — Analysis of variance and covariance

[R] **ci** — Confidence intervals for means, proportions, and counts

[R] **correlate** — Correlations (covariances) of variables or coefficients

[D] **egen** — Extensions to generate

[R] **regress** — Linear regression

[R] **predict** — Obtain predictions, residuals, etc., after estimation

[R] **regress postestimation** — Postestimation tools for regress

[R] **test** — Test linear hypotheses after estimation

[R] **summarize** — Summary statistics

[R] **table** — Tables of summary statistics

[R] **tabulate oneway** — One-way tables of frequencies

[R] **tabulate twoway** — Two-way tables of frequencies

[R] **ttest** — Mean-comparison tests

Matrices

[U] **14 Matrix expressions**

[U] **18.5 Scalars and matrices**

Mata Reference Manual

Programming

[U] **16 Do-files**

[U] **17 Ado-files**

[U] **18 Programming Stata**

[R] **ml** — Maximum likelihood estimation

Programming Reference Manual

Mata Reference Manual

System values

[R] **set** — Overview of system parameters

[P] **creturn** — Return c-class values

Internet resources

The Stata web site (http://www.stata.com) is a good place to get more information about Stata. Half of the web site is dedicated to user support. You will find answers to FAQs, ways to interact with other users, official Stata updates, and other useful information. You can also subscribe to Statalist, a listserver devoted to discussion of Stata and statistics.

You will also find information on Stata NetCourses™, which are interactive courses offered over the Internet that vary in length from a few weeks to 8 weeks. Stata also offers in-person training sessions. Visit the Stata web site for more information.

At the web site is the Stata Bookstore, which contains books that we feel may be of interest to Stata users. Each book has a brief description written by a member of our technical staff explaining why we think this book may be of interest.

We suggest that you take a quick look at the Stata web site now. You can register your copy of Stata online and request a free subscription to the *Stata News*.

Visit http://www.stata-press.com for information on books, manuals, and journals published by Stata Press. The datasets used in examples in the Stata manuals are available from the Stata Press web site.

Also visit http://www.stata-journal.com to read about the *Stata Journal*, a quarterly publication containing articles about statistics, data analysis, teaching methods, and effective use of Stata's language.

See [GSU] **19 Updating and extending Stata—Internet functionality** for details on accessing official Stata updates and free additions to Stata on the Stata web site.

19 Updating and extending Stata—Internet functionality

Internet functionality in Stata

Stata works well together with the Internet. It can use datasets and view remote help files as though they were on your computer. It can keep itself up to date (with your permission, of course). Finally, you can install *user-written commands*, which are commands that extend Stata's functionality. These are commands that have been presented in the *Stata Journal* (SJ), the *Stata Technical Bulletin* (STB), or have simply been written and shared by the greater Stata community.

This chapter will show you how you can expand Stata's horizons.

Using files from the Internet

Stata understands URLs as though they were local file locations. If you know of a file on the web that you would like to use, be it a dataset, a graph, or a do-file, you can open it in Stata easily. Here is a small example.

There are many datasets at http://www.stata-press.com/data/. Suppose that you would like to use the census12 dataset used in [U] **11 Language syntax**, and you know that its location is at http://www.stata-press.com/data/r11/census12.dta. Because you know that the command for opening a dataset is use, you could type the following:

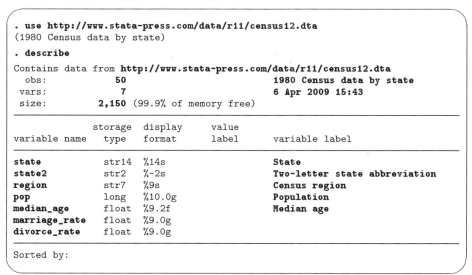

```
. use http://www.stata-press.com/data/r11/census12.dta
(1980 Census data by state)

. describe
Contains data from http://www.stata-press.com/data/r11/census12.dta
  obs:            50                          1980 Census data by state
  vars:            7                          6 Apr 2009 15:43
  size:         2,150  (99.9% of memory free)

              storage  display    value
variable name  type    format     label      variable label

state         str14   %14s                   State
state2        str2    %-2s                   Two-letter state abbreviation
region        str7    %9s                    Census region
pop           long    %10.0g                 Population
median_age    float   %9.2f                  Median age
marriage_rate float   %9.0g
divorce_rate  float   %9.0g

Sorted by:
```

This functionality is everywhere in Stata. Any command that reads a file via a *filename* in its syntax can use a web address as easily as a file that is stored on your computer.

Official Stata updates

By official Stata, we mean the pieces of Stata that are provided and supported by StataCorp. The other and equally important pieces are the user-written additions published in the SJ, distributed over Statalist, or distributed in other ways.

Stata can fetch both official updates and user-written programs from the Internet. Let's start with the official updates. StataCorp often releases updates to official Stata. These updates are to add new features and, sometimes, to fix bugs.

By default, Stata has automatic update checking turned on and set to check for updates every 7 days. We recommend using automatic update checking, because it is a simple, unobtrusive way to be sure that your copy of Stata is always up to date. If you keep this default, you will be prompted with a dialog when you start Stata if you have not checked for updates recently.

When you click on the **OK** button, a Viewer window will open that looks similar to this:

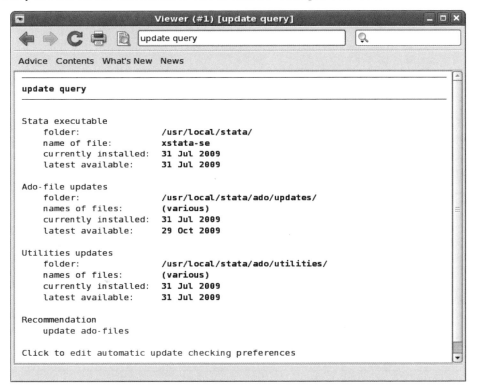

If any of the executable, ado-files, or utilities need updating, you will see a recommendation. Click on the recommendation, and Stata will update whatever needs updating.

If you need to update your executable, there is one final, crucial step: you must install your new executable. Stata makes this easy for you. Exit all instances of Stata that are running except the one in which you ran `update`. Click on the `update swap` link to complete the updating procedure. Stata will exit and start back up again.

If you would like to use Stata(console) to update Stata, you can type `update all` followed by `update swap` (if needed).

You an also check for updates at any time by choosing **Help > Official Updates** and clicking the link to http://www.stata.com. Finally, you can also type update query in the Command window. Stata will put the update results in the Results window and give you advice for what you should do next.

One troubleshooting note: If you do not have write permission for /usr/local/stata11, you cannot install official updates in this way. You may still download the official updates, but you will need to use the command-line version of update; see [U] **28 Using the Internet to keep up to date** for instructions.

Finding user-written programs by keyword

Stata has a built-in utility created specifically to search the Internet for user-written Stata programs. You can access it by selecting **Help > Search...**, choosing *Search net resources*, and entering a keyword in the field. Choosing **Help > SJ and User-written Programs** yields more specific choices for searching. The utility searches all user-written programs on the Internet, including the entire collection of *Stata Journal* and STB programs. The results are displayed in the Viewer, and you can click to go to any of the matches found.

For the syntax on how to use the equivalent search *keywords*, net command, see [R] **search**.

Downloading user-written programs

Downloading user-written programs is easy. Start by selecting **Help > SJ and User-written Programs**:

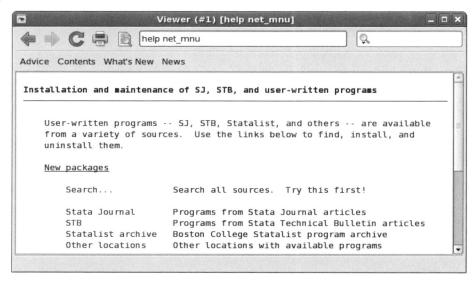

As the Viewer says, try Search... first.

Suppose that you were interested in finding more information or some user-written programs involving cubic splines. You select **Help > Search...**, select *Search all*, type cubic spline in the search box, and click on the **OK** button.

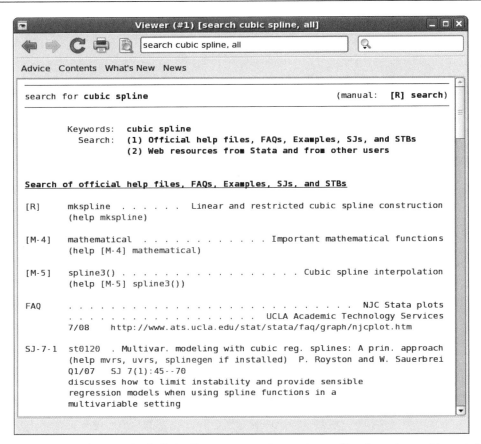

The first entry points to the built-in Stata command mkspline. You investigate this and find it interesting. You see that the next two entries point to some built-in routines in Mata. You follow these links, because Mata is not only intriguing but also fast. You see the next link points to an FAQ on Stata's web site. Finally, you decide to check the fifth link. It points to an article in the *Stata Journal*, volume 7, number 1 (first quarter, 2007). You should click on the st0120 link, because this will go to the programs associated with this article.

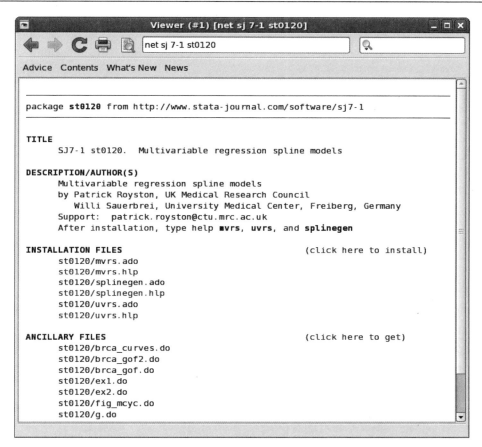

There are many files here! You can click on the help files to see if the commands here look like they are interesting. If you decide that you would like to try the commands, you can click on the link `click here to install`. If you decide that you would like to use some of the ancillary files—files that typically help explain the workings of the command, you could install those, too. You do not need to worry—doing so will not interfere in any way with your copy of Stata. We will show you how to uninstall these programs safely shortly. Click on the install link to install the package. That is all there is to installing a user-written package.

Now suppose that you decide that you would like to uninstall the package. Doing so is simple enough: select **Help > SJ and User-written programs**, and click on the `List` link. You should see the following:

(Continued on next page)

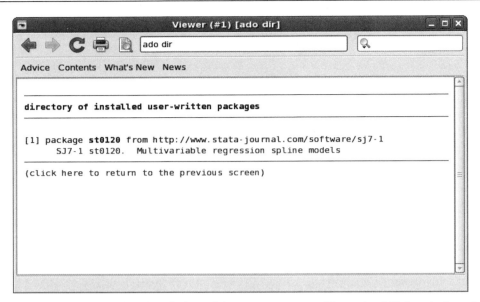

If you click on the one-line description of the program, you will see the full description of what has been installed. Here is a short version of what you would see:

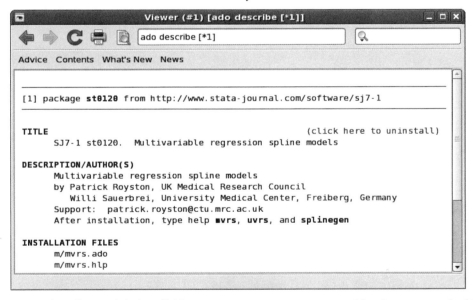

You can uninstall materials by clicking on click here to uninstall when you are looking at the package description. Try it.

For information on downloading user-written programs by using the net command, see [R] **net**.

A Troubleshooting Stata

Contents

A.1 If Stata(console) and Stata(GUI) do not start

If you have attempted to invoke Stata(GUI) by typing `xstata`, `xstata-se`, or `xtstata-mp` at the Unix prompt and it has failed, attempt to invoke Stata(console) by typing `stata`, `stata-se`, `stata-mp` at the Unix prompt. If Stata(console) fails, continue here. If Stata(console) starts without problems, see the next section of this chapter.

You tried to start Stata and it refused; Stata or your operating system presented a message explaining that something is wrong. Here are the possibilities:

cannot find Stata directory
Stata first checks in `/usr/local/stata11` and then checks in `/usr/local/stata` to find the license file. If Stata does not find the license in either of these locations, it looks in every directory in the Unix path. If you receive this message, most likely the directory where Stata is installed is not in your path. You need to add this directory to your path.

Cannot find license file
This message means just what it says; nothing is too seriously wrong. Stata simply could not find what it is looking for. The two most common reasons for this are that you did not complete the installation process or Stata is not installed where it should be.

Did you insert the codes printed on your paper license to unlock Stata? If not, go back and complete the initialization procedure.

Assuming that you did unlock Stata, Stata is merely mislocated, or the location has not been filled in.

Error opening or reading the file
Something is distinctly wrong and for purely technical reasons. Stata found the file that it was looking for, but either the operating system refused to let Stata open it or there was an I/O error.

The `stata.lic` file could have incorrect permissions. Verify that `stata.lic` is in the directory where Stata is installed, likely `/usr/local/stata11` and that everybody has been granted read permission. To change the permissions, become superuser by logging in as root, using `su`, or `sudo`, and type `chmod a+r /usr/local/stata11/stata.lic`.

Other messages
The other messages indicate that Stata thinks you are attempting to do something that you are not licensed to do. Most commonly, you are attempting to run Stata over a network when you do not have a network license, but there are many other alternatives. There are two possibilities: either you really are attempting to do something that you are not licensed to do, or Stata is wrong. In either case, you are going to have to contact us. Your license can be upgraded, or, if Stata is wrong, we can provide codes to make Stata stop thinking that you are violating the license; see [U] **3.9 Technical support**.

A.2 If Stata(console) starts but Stata(GUI) does not

Trouble with libraries

Stata, like many other programs designed for Unix machines, needs certain system libraries to run. Most of the standard library routines that Stata needs are included in the Stata binary. However, Stata does rely on a few external libraries. For example, Stata assumes that your system will have the standard C library available. Stata(GUI) assumes that you have the X Windows libraries as well. These libraries are often located in different places on different implementations of Unix. For example, under Linux, the standard C libraries can be found in /lib, whereas the X Windows libraries are in /usr/X11R6/lib; under Solaris, the standard libraries are in /usr/lib, whereas X libraries are in /usr/openwin/lib. The operating system needs to know where to find the libraries that Stata needs to run. The Unix environment variable LD_LIBRARY_PATH tells the operating system where to find libraries. If you get an error message that says something like

```
ld.so.1: xstata: fatal: some_library.so.#: can't open file
```

the likely reason is that the operating system cannot find the necessary libraries. Stata does not rely on any unique libraries; you can rest assured that the libraries you need are on your system. You should look for the library in question on your system and make sure the environment variable LD_LIBRARY_PATH includes the path to the directory where the library in question is located. Your system administrator may be able to help with this task.

Setting the DISPLAY environment variable

Sometimes when executing Stata in a networked environment, Stata(GUI) will produce the error message

```
You need X Windows for this version of Stata.
```

This means that Stata has not found the DISPLAY environment variable. You need to set the DISPLAY variable to be the screen on which you want Stata(GUI) to appear. For example, in csh, type setenv DISPLAY machine:0.0.

xhost permissions

Another related error message can occur with Stata(GUI). When Stata is being run in a networked environment, the computer on which Stata(GUI) is actually executing may not have permission to draw on the screen of the computer on which you have asked Stata to draw. Then, you may see the following:

```
Xlib: connection to machine_name:0.0 refused by server
Xlib: Client is not authorized to connect to Server
xhost: unable to open display
```

This means that the machine that is actually executing the Stata instructions (the x-client in X Windows parlance) does not have permission to draw on the screen of the computer that you have asked it to (the x-server). On the machine on which you want to display Stata(GUI), type

```
% xhost +client_machine
```

This will give the client permission to draw on the server.

Getting more help

If you continue to experience problems invoking Stata(GUI), please see the Unix FAQs on our web site at http://www.stata.com/support/faqs/unix/, or contact Stata Technical Support.

A.3 Troubleshooting tips

If you had trouble with the installation of Stata, it could be because you have a recent copy of Unix that will not allow you to run applications from a DVD drive. Type `df -l` to see what local devices are mounted; one should look like the Stata DVD. For example, you could see something like

```
$ df -l
Filesystem              1K-blocks      Used Available Use% Mounted on
/dev/hda6                23054660   5528268  16336380  26% /
/dev/hdc                   274158    274158         0 100% /media/Stata
```

`/dev/hdc` is the device name, whereas the `/media/Stata` is the mount point.

If you see something indicating that the Stata DVD is successfully mounted, you need to see if you are being prevented from running applications directly from the DVD. Type `mount` to get information about any mounted file systems. Somewhere on the list, you should see information about your DVD. Continuing with the above example, you should see your device name and mount point in the output:

```
$ mount
  omitted output
/dev/hdc on /media/Stata type iso9660 (ro,noexec,nosuid,nodev,uid=220) [STATA]
```

If you see the term `noexec` appear, you are not allowed to run applications from a DVD. Your best course of action is to copy everything from the Stata DVD to a temporary directory and run the installation from there. Substitute your mount point for `/media/Stata` below.

```
$ mkdir /tmp/statainstall
$ cp -r /media/Stata /tmp/statainstall
$ mkdir /usr/local/stata11
$ cd /usr/local/stata11
$ sudo /tmp/statainstall/install
```

After you have Stata running and initialized, you can delete the temporary directory `/tmp/statainstall`.

If you are still having problems installing or if you have any other troubles, please see the Unix FAQs on our web site at http://www.stata.com/support/faqs/unix/. If this does not help, contact Stata Technical Support. Please gather all the information you can about your system, including your computer model and the type and version of Unix that you are using. Finally, we need your Stata serial number and the date your version of Stata was "born". Include them if you email, and know them if you call. You can obtain them by typing `about` in Stata's Command window. `about` lets us know everything about your copy of Stata, including the version and the date it was produced.

Notes

B Managing memory

Contents

B.1 Memory size considerations

Stata works with a copy of the dataset that it loads into memory.

By default, Stata/IC allocates 10 MB to Stata's data areas, and you can change it.

By default, Stata/MP and Stata/SE allocate 50 MB to Stata's data areas, and you can change it.

You can even change the allocation to be larger than the physical amount of memory on your computer because Unix provides virtual memory.

Virtual memory is slow but adequate in rare cases when you have a dataset that is too large to load into real memory. If you use large datasets often, we recommend that you add more memory to your computer. See [GSU] **B.4 Virtual memory and speed considerations**.

You can change the allocation when you start Stata, as will be discussed in [GSU] **C.2 Specifying the amount of memory allocated**.

You can also change the total amount of memory allocated while Stata is running and optionally make that setting the default to be used by future invocations of Stata. That is the topic of this appendix.

It does not much matter which method you use. Being able to change the total on the fly is convenient, but even if you cannot do this, it just means that you must specify it ahead of time, and if later you need more, you must exit Stata and reinvoke it with the larger total.

Do not specify more memory than you actually need. Doing so does not benefit Stata; in fact, it may slow Stata down!

B.2 Setting the size on the fly

Assume that you have changed nothing about how Stata starts, so you have the default amount of memory (50 MB for Stata/MP and Stata/SE, 10 MB for Stata/IC) allocated to Stata's data areas. You are working with a large dataset and now wish to increase memory to 100 MB. You can type

```
. set memory 100m
```

and, if your operating system can provide the memory to Stata, Stata will work with the new total. Later in the session, if you want to release that memory and work with only 2 MB, you could type

```
. set memory 2m
```

There is only one restriction on the set memory command: whenever you change the total, there cannot be any data already in memory. If you have a dataset in memory, you save it, clear memory, reset the total, and then use it again. We are getting ahead of ourselves, but you might type

```
. save mydata, replace
file mydata.dta saved
. clear
. set memory 100m
...
. use mydata
```

When you request the new allocation, your operating system might refuse to provide it:

```
. set memory 1800m
op. sys. refuses to provide memory
r(909);
```

If that happens, you are going to have to take the matter up with your operating system. In the above example, Stata asked for 1800 MB, and the operating system said no.

B.3 The memory command

memory helps you figure out whether you have enough memory to do something. If you are using Stata/SE (or Stata/MP), you could type memory:

```
. memory
```

	bytes	
Details of **set memory** usage		
overhead (pointers)	114,136	10.88%
data	913,088	87.08%
data + overhead	1,027,224	97.96%
free	21,344	2.04%
Total allocated	1,048,568	100.00%
Other memory usage		
set maxvar usage	1,816,666	
set matsize usage	1,315,200	
programs, saved results, etc.	772	
Total	3,132,638	
Grand total	4,181,206	

You can type memory in Stata/IC, too, but the output will vary slightly from that of Stata/SE and Stata/MP; see [D] **memory**. Having 21,344 bytes free is not much. You might increase the amount of memory allocated to Stata's data areas by specifying set memory 2m.

```
. save nlswork
. clear
. set memory 2m
...
. use nlswork
(National Longitudinal Survey.  Young Women 14-26 years of age in 1968)
```

```
. memory
```

	bytes	
Details of **set memory** usage		
overhead (pointers)	114,136	5.44%
data	913,088	43.54%
data + overhead	1,027,224	48.98%
free	1,069,920	51.02%
Total allocated	2,097,144	100.00%
Other memory usage		
set maxvar usage	1,816,666	
set matsize usage	1,315,200	
programs, saved results, etc.	773	
Total	3,132,639	
Grand total	5,229,783	

More than 1 MB is now free; that's better. See [D] **memory** for more information.

You can use `set memory` to set the memory for future Stata sessions, as well as the current session, by using the `permanently` option. If, for example, you wanted 100 MB of memory allocated for data in future Stata sessions, you could type

```
. set memory 100m, permanently
```

B.4 Virtual memory and speed considerations

When you open (`use`) a dataset in Stata for Unix, Stata loads the entire dataset into memory.

When you use more memory than is physically available on your computer, Stata slows down. If you are using only a little more memory than is on your computer, performance is probably not too bad. On the other hand, when you are using a lot more memory than is on your computer, performance will be noticeably affected. We recommend then that you type

```
. set virtual on
```

Virtual memory systems exploit locality of reference, which means that keeping objects closer together allows virtual memory systems to run faster. `set virtual` controls whether Stata should perform extra work to arrange its memory to keep objects close together. By default, `virtual` is set `off`.

In general, you want to leave `set virtual` set to the default of `off` so that Stata will run faster.

When you `set virtual on`, you are asking Stata to arrange its memory so that objects are kept closer together. This setting requires Stata to do a substantial amount of work. We recommend setting virtual on only when the amount of memory in use drastically exceeds what is physically available. In these cases, setting virtual on will help, but keep in mind that performance will still be slow. See [U] **6.5 Virtual memory and speed considerations**. If you are using virtual memory often, you should consider adding memory to your computer.

Notes

C Advanced Stata usage

Contents

C.1 Executing commands every time Stata is started

Stata looks for the file `profile.do` when it is invoked and, if it finds it, executes the commands in it. Stata looks for `profile.do` first in the directory where Stata is installed, then in the current directory, then along your path, and finally along the ado-path (see [P] **sysdir**); we recommend that you put `profile.do` in your bin directory `$HOME/bin`.

Say that every time you start Stata you want `matsize` set to 800 (see [R] **matsize**). Create the file `profile.do` in `$HOME/bin` containing

```
set matsize 800
```

When you invoke Stata, this command will be executed:

```
(usual opening appears, but with the addition)
running /home/mydir/bin/profile.do ...
. _
```

You could also type `set matsize 800, permanently` in Stata. The `permanently` option tells Stata to remember the setting for future sessions and eliminates the need to put the `set matsize` command in `profile.do`.

`profile.do` is treated just as any other do-file once it is executed; results are just the same as if you had started Stata and then typed `run profile.do`. The only special thing about `profile.do` is that Stata looks for it and runs it automatically.

System administrators may also find `sysprofile.do` useful. This file is handled in the same way as `profile.do`, except that Stata looks for `sysprofile.do` first. If that file is found, Stata will execute any commands it contains. After that, Stata will look for `profile.do` and, if that file is found, execute the commands in it.

One example of how `sysprofile.do` might be useful is for system administrators who want to change the path to one of Stata's system directories. Here `sysprofile.do` could be created to contain the command

```
sysdir set SITE "/opt/stata/ado"
```

See [U] **16 Do-files** for an explanation of do-files. They are nothing more than ASCII text files containing a sequence of commands for Stata to execute.

C.2 Specifying the amount of memory allocated

You can change the amount of memory once Stata is running by using the set memory command, and you can make that setting permanent; see [GSU] **B Managing memory**.

C.3 Advanced starting of Stata for Unix

The syntax of the command to start Stata(GUI) is

$$\text{xstata} \ \big[\text{-}option \ \big[\text{-}option \ \big[\dots\big]\big]\big] \ \big[stata\_command\big]$$

The syntax of the command to start Stata(console) is

$$\text{stata} \ \big[\text{-}option \ \big[\text{-}option \ \big[\dots\big]\big]\big] \ \big[stata\_command\big]$$

If you have Stata/SE, the commands are xstata-se and stata-se. If you have Stata/MP, the commands are xstata-mp and stata-mp.

The allowable options are

-h	display usage diagram
-q	suppress logo and initialization messages
-m#	amount of memory to allocate; default is -m10 for Stata/IC and -m50 for Stata/MP and Stata/SE
-b	set background (batch) mode and log in ASCII text (console only)
-s	set background (batch) mode and log in SMCL (console only)

Typing stata -h does not start Stata but just shows the syntax diagram for invoking Stata.

The -q option starts Stata, but it suppresses all the initialization messages, including the Stata logo.

The -m option specifies the amount of memory to be allocated to Stata's data areas; this is discussed in [GSU] **C.2 Specifying the amount of memory allocated**. Most users find typing set memory after Stata launched more convenient; see [U] **6 Setting the size of memory**.

The -b and -s options specify batch mode; see [GSU] **C.4 Stata batch mode**.

C.4 Stata batch mode

If you want to use Stata in batch mode, we suggest using Stata(console). Typing

```
% stata -s do bigjob
```

tells Stata to execute the commands in bigjob.do, suppress all screen output, and route the output to bigjob.smcl in the same directory.

```
% stata -b do bigjob
```

tells Stata to execute the commands in bigjob.do, suppress all screen output, and route the output to bigjob.log in the same directory.

You can also run the above examples in the background by typing

```
% stata -s do bigjob &
% stata -b do bigjob &
```

You may also use redirection:

```
% stata < bigjob.do > bigjob.log &
```

Warning: if your do-file contains either the #delimit command or comment delimiters (/* and */ or ///), the second method will not work. Also, the second method cannot produce SMCL output. We recommend that you use the first methods: stata -s do bigjob & or stata -b do bigjob &.

Note: Stata runs profile.do before doing bigjob.do, just as it would interactively.

C.5 Using X Windows remotely

Suppose that you are sitting in front of a computer named local and that you wish to run Stata on a computer named neighbor.

1. Tell X Windows on local to allow neighbor to use its display by typing xhost +neighbor.

2. Be sure that you have set X Windows' DISPLAY environment variable on neighbor to contain local:0.0. X requires this. Important: this variable must be set on neighbor.

Having done this, Stata should work:

```
local% xhost +neighbor
local% ssh neighbor
neighbor% setenv DISPLAY local:0.0
neighbor% xstata
```

At this point either Stata launches, or you see

```
Xlib:  connection to "local:0.0" refused by server
Xlib:  Client is not authorized to connect to server
```

Here you will have to get help from your network administrator. Proper authorizations have not been given, and these problems have nothing to do with Stata.

To make this process simpler, you may also be able to use the -X flag when invoking ssh:

```
local% ssh -X neighbor
neighbor% xstata
```

Whether this works depends on your Unix installations. The first series of commands should always work.

C.6 Summary of environment variables

Environment variable	Description
HOME	User's home directory. Default is the directory specified in /etc/passwd.
PATH	Unix executable search path.
SHELL	What to execute when users try to shell out of Stata. Default is /bin/sh.
S_ADO	Sets Stata's ado-path.
STATATERM	Used only if you want to use a different termcap or terminfo entry from what is in your TERM environment variable.
STATATMP	Sets Stata's temporary directory. Default is /tmp.

Notes

D Stata manual pages for Unix

Contents

(Continued on next page)

139

Title

conren — Set the color, etc., of Stata(console)

Syntax

High-level commands

conren

conren style #

conren ul #

conren test

conren clear

Low-level commands

set conren

set conren clear

set conren [sf | bf | it] { result | { txt | text } | input | error |

    link | hilite } [char [char ...]]

set conren { ulon | uloff } [char [char ...]]

set conren reset [char [char ...]]

set conren off [char [char ...]]

where *char* is

{ *any_character* | < # > }

Note

This command concerns Stata for Unix only and, in particular, the Stata you launch by typing stata or stata-se, not xstata or xstata-se, also known as Stata(console) or the non-GUI version of Stata.

Description

conren and set conren may improve display of the output on your screen. Some terminals, for instance, can display colors, but Stata may not know that your terminal has that capability. Some terminals have multiple intensities or boldfaces. Some terminals can underline. The high-level conren command lets you set a display style and/or underlining scheme from among a selection of predefined settings.

conren `style` followed by a scheme number sets color and font codes on the basis of the underlying scheme.

conren `ul` followed by an underlining scheme number sets the codes allowing underlining.

conren with no arguments displays a message explaining the command and telling the range of style and underlining scheme numbers available.

conren `test` displays three columns of output in sf (standard face) font, **bf (boldface) font**, and *it (italics) font* showing the assignment of colors with and without underlining.

conren `clear` clears all the currently defined display style and underlining definitions.

The low-level `set conren` command lets you view and set specific display and underlining codes.

`set conren` displays a list of the currently defined display codes.

`set conren clear` clears all codes.

`set conren` followed by a font type (`bf`, `sf`, or `it`) and display context (`result`, `text`, `input`, `error`, `link`, or `hilite`) and then followed by a series of space-separated characters sets the code for the specified font type and display context. If the font type is omitted, the code is set to the same specified code for all three font types.

`set conren ulon` and `set conren uloff` set the codes for turning underlining on and off.

`set conren reset` sets the code that will turn off all display and underlining codes.

`set conren off` sets the code used by Stata when it exits and returns control back to the operating system.

When Stata launches, it is as if you have typed

        . conren clear

which is equivalent to the low-level command

        . set conren clear

meaning that Stata is to assume that your monitor cannot display different colors, intensities, or underlining. Stata makes this assumption because, were Stata to assume otherwise and your terminal could not provide the capability, the result could look really bad. Thus, a few minutes of playing around can be well worth the effort, and you do not have to be a computer expert to do this. You cannot hurt anything permanently by typing the wrong command.

The next-to-worst thing that can happen is that Stata's output will look so bad that you cannot even read it, and then just exit Stata. Stata will be fine the next time.

The worst thing that can happen is that your window/screen/terminal will look so garbled that you will have to close it and open a new one (or, if it really is a separate terminal, turn it off and turn it back on).

Once you are happy with your settings, you can put the `set conren` commands in your `profile.do` so that they are executed at the start of every Stata session.

(*Continued on next page*)

Finding a color scheme

First, let's try various color schemes. What will work and look good depends on your terminal/monitor and whether you are using a white or black background. (We really prefer a black background for Stata, and if you are using a white background, we recommend that you try black someday. We prefer a black background for Stata since, by default, it uses green and yellow for most output, and these colors do not show up well on a light-colored background. Switching the background color, however, is something that you will have to take up with Unix, not Stata.)

First, type the following:

```
. conren
```

Doing so first tells you the number of possible display schemes and underlining schemes available. There are a few underlining schemes and many more display schemes. Some of these schemes were designed with black backgrounds in mind, whereas others were designed for white backgrounds. We suggest that you first select a display style scheme, and then after finding the scheme you like, explore the possible underlining schemes.

You would type

```
. conren style 1
```

to try display style scheme 1. `conren style` and `conren ul` automatically run `conren test` so that you can see the result on your screen. If the result is obviously bad, move on and try another scheme. If the resulting color scheme might be reasonable, try out Stata and see what you think. Try several commands, and look at a few help files to see if the selected display style scheme is appropriate. You can always return to the default with

```
. conren clear
```

which may be hard if you cannot even see what you are typing. Remember, if things are really bad, just type `exit` and then restart Stata.

Try all the prerecorded schemes to determine which you like best.

Can your terminal underline?

Type `conren test` to look at the various output types. Is the word `underlined` underlined—the underlining being on the same line and actually touching the characters—or is it instead more crudely rendered with a string of dashes underneath, on a second line?

If the word `underlined` is underlined, skip this section; evidently Stata has already figured out that your terminal can underline and is doing that.

Sometimes, Stata cannot figure that out for itself. Let's see if you can underline. Type

```
. conren ul 1
```

Now look at the output from `conren test` again. Is `underlined` underlined or just a mess? If it is a mess, you can remove the underlining codes (while leaving the display style codes untouched) by typing

```
. conren ul 0
```

You can now try the other available underlining schemes to see if they work any better for you.

If you had success...

So, let's say that you discovered that what works best for you is

```
. conren style 4
. conren ul 1
```

and you just had no success with boldface at all. The next time you enter Stata, if you want the prettier look, you will have to type those two commands. To avoid that, create a file `profile.do` and put those two lines in that file. Actually, we suggest that you put the lines in the file as

```
quietly conren style 4
quietly conren ul 1
```

because, if you also use Stata in batch mode, using the `quietly` option will prevent odd messages from appearing when Stata starts.

If you did not have success...

Well, now you really need to be technical. It is possible to make Stata's output look prettier if you know the "escape sequences" to cause special effects on your terminal.

Pretend that the codes for your terminal to turn underlining on and off were Esc-[4m and Esc-[24m. You could tell Stata that by typing

```
. set conren ulon <27> [ 4 m
. set conren uloff <27> [ 2 4 m
```

Escape has the decimal code 27, and you can type decimal codes by enclosing them in less-than and greater-than signs. Regular characters you can just type. Remember, however, that you must type at least one space between each character.

All the features can be set in this way. If you type

```
. set conren
```

Stata will report what is currently set.

Also see

[P] **smcl** — Stata Markup and Control Language

Title

stata — Stata invocation command

Syntax

xstata-mp [*-option* [*-option* [...]]] [*stata_command*]

xstata-se [*-option* [*-option* [...]]] [*stata_command*]

xstata [*-option* [*-option* [...]]] [*stata_command*]

stata-mp [*-option* [*-option* [...]]] [*stata_command*]

stata-se [*-option* [*-option* [...]]] [*stata_command*]

stata [*-option* [*-option* [...]]] [*stata_command*]

where the *options* are

-h   display usage diagram
-q   suppress logo and initialization messages
-m#   amount of memory to allocate; default is -m10 for
          Stata/IC and -m50 for Stata/MP and Stata/SE
-s   set background (batch) mode and log in SMCL (console only)
-b   set background (batch) mode and log in ASCII text (console only)

Description

xstata-mp starts the GUI version of Stata/MP. xstata-se starts the GUI version of Stata/SE. xstata starts the GUI version of Stata/IC.

stata-mp starts the console version of Stata/MP. stata-se starts the console version of Stata/SE. stata starts the console version of Stata/IC.

Remarks

Here are the explanations for the startup options.

Typing stata -h does not enter Stata but instead shows the syntax diagram for invoking Stata.

The -q option suppresses all the initialization messages including the Stata logo when Stata is invoked.

The -m option controls the amount of memory allocated by Stata at initialization.

The -s and -b options are for users who do not wish to run Stata interactively. Typing

```
% stata -s do bigjob
```

tells Stata to execute the commands in bigjob.do, suppress all screen output, and route the output to bigjob.smcl. Typing

```
% stata -b do bigjob
```

tells Stata to execute the commands in `bigjob.do`, suppress all screen output, and route the output to `bigjob.log`.

The previous two commands are almost equivalent to

```
% stata < bigjob.do > bigjob.log
```

Warning: if your do-file contains either the `#delimit` command or comment delimiters (`/*` and `*/` or `///`), the second method will not work. Also, the second method cannot produce SMCL output. We recommend that you use the first method: `stata -s do bigjob` or `stata -b do bigjob`.

If you do use the first method, bear in mind that Stata automatically begins a log. If the *stata_command* is of the form

$$\{ \mathrm{do} \mid \mathrm{run} \} \; \big[\, path \,\big] \, filename \big[\, .suffix \,\big]$$

the output is routed to *filename*.`smcl` in the current directory if you used `-s` and to *filename*.`log` in the current directory if you used `-b`. Otherwise, the output is routed to `stata.smcl` or `stata.log`. In either case, if the log file already exists, it is silently erased before starting.

Whether you use the first or second method, whenever Stata is running without a terminal, it sets `c(mode)` to contain `batch`. If Stata is running in interactive mode, `c(mode)` is set to contain " ". See [P] **creturn**.

❑ Technical note

Many users log in to Unix computers from Windows or Mac computers. If you are one of these users, there are several third-party packages that allow you to display X Windows graphics on Windows or Mac computers.

❑

Also see

[U] **6 Setting the size of memory**

Subject index

This is the subject index for the *Getting Started with Stata for Unix* manual. Readers may also want to consult the combined subject index (and the combined author index) in the *Quick Reference and Index*. The combined index indexes the *Getting Started* manuals, the *User's Guide*, and all the reference manuals, except the *Mata Reference Manual*.